Sustainable Ocean Resource Governance

Deep Sea Mining, Marine Energy and Submarine Cables

Edited by

Markus Kotzur
Nele Matz-Lück
Alexander Proelss
Roda Verheyen
Joachim Sanden†

BRILL
NIJHOFF

LEIDEN | BOSTON

Library of Congress Cataloging-in-Publication Data

Names: Kotzur, Markus, editor.
Title: Sustainable ocean resource governance : deep sea mining, marine energy, and submarine cables / edited by Markus Kotzur, Nele Matz-Lück, Alexander Proelss, Roda Verheyen, Joachim Sanden.
Description: Leiden ; Boston : Brill/Nijhoff, [2018] | Includes bibliographical references and index.
Identifiers: LCCN 2017054586 (print) | LCCN 2017055327 (ebook) | ISBN 9789004360273 (E-book) | ISBN 9789004360266 (hardback : alk. paper)
Subjects: LCSH: Marine resources development—Law and legislation. | Renewable energy sources—Law and legislation. | Sustainable development—Law and legislation. | Ocean mining—Law and legislation. | Ocean bottom—Law and legislation. | OSPAR Commission. | Cables, Submarine—Law and legislation.
Classification: LCC K3485 (ebook) | LCC K3485 .S87 2018 (print) | DDC 341.4/5—dc23
LC record available at https://lccn.loc.gov/2017054586

Typeface for the Latin, Greek, and Cyrillic scripts: "Brill". See and download: brill.com/brill-typeface.

ISBN 978-90-04-36026-6 (hardback)
ISBN 978-90-04-36027-3 (e-book)

This book is printed on acid-free paper and produced in a sustainable manner.

Contents

PART 2
Sustainable Ocean Resource Governance – The Specific Fields of Application

Foreword: Really a Sea Change? – In Search of a Coherent and Consistent Sustainability Approach

The title originally chosen for the invitation to the third Hamburg International Environmental Law Conference (HIELC 2016, April 15 and 16, 2016) is, admittedly, not self-explanatory: "A Sea Change for Sustainable Ocean Resource Governance". It might both require a question mark and deserve an exclamation mark when alluding to the notion of a "sea change". The famous phrase stems from W. Shakespeare's tragic comedy *The Tempest* which puts the sorcerer Prospero, in fact the usurped Duke of Milan, on a remote island, where he uses illusions and skilful manipulations such as the eponymous tempest to re-establish his daughter Miranda's rightful position in society. It is Prospero's magical servant, the spirit Ariel, who declaims: "Full fathom five thy father lies: Of his bones are coral made: Those are pearls that were his eyes: Nothing of him that doth fade but doth suffer a sea-change into something rich and strange".

Quite obviously, "sea change" is used as an emblematic idiom for a profound transformation of something into something else. To discuss such a profound transformation in the way humankind makes sustainable use of maritime resources, reconciles economical with ecological interests and treats the oceans as truly its common heritage, can indeed be seen as an ambitious endeavour of Shakespearian dimensions. As *Vladimir V. Golitsyn*, President of the Hamburg based International Tribunal for the Law of the Sea (ITLOS), stated in his opening address, a sea change "implies a shift in the international community's treatment of marine resources" – a shift away from the "laissez-faire-approach" still valid "until as late as the mid-twentieth century" to "sound management and conservation". Notwithstanding the often expressed State consensus that the sustainable use and exploitation of the oceans' abundant resources requires effective ocean governance, many sceptics – or should we call them realists? – are concerned that without a sorcerer's magic powers the "shift" mentioned by President *Golitsyn* is doomed to failure. Sorcery and science, enchantment and politics, undoubtedly tread very different paths. They have very different instruments at their command to turn the world into a different place. The magic words' immediate success might be the comforting illusion; the scientists' achievements might rather be the discovery of a discomforting reality. The latter, however, can bring about a burdensome and lengthy but in the long run promising up-hill struggle towards a better outcome. Realism meets idealism, so to speak.

The landmark Paris climate conference of November/December 2015 displayed political will of the more idealistic kind. Following the adoption of the United Nations Framework Convention on Climate Change in 1992, 195 States ultimately adopted what could be described as the second global climate deal being legally binding in its application and universal in its regulatory aims. The USA's intended withdrawal in 2019 is doubtlessly a return to harsh reality and a severe setback for the aspirational project but, as reassured by 19 out of 20 participating states during the Hamburg G20 summit in July 2017, not the end of the common climate-protective route embarked in Paris. Among others, the new Paris Agreement includes the long-desired and finally express recognition for the oceans as crucial factors in any climate regulation scheme. Under the programmatic category of "Ecosystem Integrity", the marine ecosystems are mentioned both within the Preamble and within the operative part of the Treaty. The signatories expressly note "the importance of ensuring the integrity of all ecosystems, including oceans". They furthermore subscribe to the "sustainable management of natural resources" and promise, in Article 5, to "take action to conserve and enhance, as appropriate, sinks and reservoirs of greenhouse gases".

HIELC 2016 – the outcomes of which we are happy to present and put forward for critical discussion in the volume at hand – not only endorses the "realism meets idealism" leitmotiv, but also follows up the vision of solely making sustainable and eco-friendly use of natural resources. In doing so, it addresses three key issues at the interface between sustainable economic growth, effective resource management and urgent environmental protection: deep sea mining, marine energy generation, and offshore pipeline and cable systems. It has been proven by experience, however, that sustainable environmental protection and sustainable economic growth more often appear as foes than friends. The sectoral nature of current ocean governance and the existing patchwork of management arrangements for the oceans do not facilitate the search for a coherent and consistent sustainability approach. This became quite clear when President *Golitsyn* in his opening remarks addressed the wider horizons of sustainable ocean resource governance and referred to the relevant statements made by the ITLOS, in particular in the 2011 Advisory Opinion of the Tribunal's Seabed Disputes Chamber.

With regard to marine energy generation, these wider horizons were first substantiated by *Gabriele Goettsche-Wanli*, who provided a global perspective on sustainable production of offshore renewable energy, followed by *Henning Jessen's* comment on the same topic. *G. Goettsche-Wanli* called for an "enabling legal environment" and foresaw effective offshore energy governance based on cross-sectoral collaboration as well as stakeholder participation.

H. Jessen agreed to the necessity of an integrated, interdisciplinary and intersectoral approach. He reminded the States of their responsibilities by referring to them as "stewards of the global marine environment". *Jessen*'s metaphor recalls – maybe not accidentally – *Eyal Benvenisti's* notion of "Sovereigns as Trustees of Humanity" when analysing the "Accountability of States to Foreign Stakeholders" (in: AJIL 107 (2012), pp. 295). The various types of energy from the ocean and in the oceans, as well as emerging technologies for deriving electrical power from the ocean (such as tidal power, wave power, ocean thermal energy conversion, etc.), indeed address the interests of humanity as such and open new perspectives for renewable energy creation. Fossil fuels – petroleum or natural gas underneath the ocean floor – may also be seen as some form of – though not renewable – ocean energy. The use and exploitation of all forms of marine energy is, however, highly environment-sensitive and needs – both offshore and onshore – to respect up-to-date standards of environmental protection.

How difficult it still is to establish an overall legal framework for the sustainable use of marine genetic resources became quite clear when *Rüdiger Wolfrum* shared his reflections on the "Realization of Sustainable Management/ Development under the Law of the Sea Convention?" – a Convention that has been labelled by the President of the Third United Nations Conference on the Law of the Sea, *Tommy Koh*, as the "Constitution for the oceans". *Wolfrum* emphasised that the principle of sustainability became "the leitmotiv of international environmental law" and endorsed the idea "that natural resources are managed in a way that they may benefit future generations" as well-established in international law and continuing to gain further ground. A more particular perspective was taken by *David Kenneth Leary* who discussed "International Environmental Law, Sustainable Generation of Energy from the Ocean and Small Island Developing States in the Pacific". *Yoshifumi Tanaka* and *David Freestone* finally developed paths towards a "Sustainable Management of Marine Natural Resources". Such a sustainability approach ought to take into account social development, economic growth and environmental protection, with particular emphasis on the needs and interests of future generations. The transformation of efficient exploitation into sustainable exploitation is thus inevitable.

It is not the intention of this foreword to comprehensively summarize the conference's topics and to comment on every single submission of our esteemed authors. They speak for themselves. That said, it shall be highlighted here that the two above-mentioned keynotes and the relevant comments opened the floor for discussions on specific fields of application of sustainable ocean resource governance. In addition to marine energy generation, they encompass deep sea mining, as well as submarine cable systems. Concerning deep sea mining, while advocates believe that the future of mining lies at the bottom of the seas, opponents fear that industry-driven interests will do anything but provide adequate solutions to the urgent ocean sustainability challenges. The relatively new technique of *"deep sea mining"* indeed raises concerns regarding its potential impact on marine and coastal ecosystems. Since the complete consequences of full-scale mining operations are still subject to scientific uncertainty and further research is therefore necessary to fully understand all potential risks, the law governing this field has to qualify as a blueprint for effective risk management. The same holds true for providing legal instruments governing the development and protection of submarine cable systems. The deep seabed, being mankind's common heritage, forms the habitat for the world's underwater networks which needs to be governed by law. We greatly appreciate the most insightful contributions by *Douglas R. Burnett, Edwin Egede, Pablo Ferrara*, and *Keyuan Zou* on these issues. They also address questions of liability and responsibility, situating their arguments in the overall framework of the "common heritage of mankind". The views expressed in all articles are those of the authors and should not be taken as reflecting in any way the policies or views of the institutions they are representing.

To conclude, it was a great pleasure for us to host HIELC 2016 in the Free and Hanseatic City of Hamburg. We are honoured to continue the conference scheme established by our colleagues *Hans-Joachim Koch* (who retired) and *Doris König* (who has been appointed as judge of the German Federal Constitutional Court). We are glad that the conference brought together some 120 participants from nearly 30 countries worldwide. Only such a broad appearance makes it possible to successfully discuss global environmental problems and correlate national and international legal measures – a chance for a "sea change" through scientific exchange and inter-cultural communication.

We would like to thank all who helped to make HIELC 2016 a success and in particular:

- the City of Hamburg and First Mayor Olaf Scholz for enabling Internationaler Umweltrechtstag Hamburg e.V. to organise the conference,
- all our sponsors for their generous support,
- Christina Simmig and her student assistant Julia Bialek for the extraordinary job they did in managing the event; C. Simmig has been supporting the conference from preparation until publishing this 3rd HIELC-volume and has been a reliable contact person for nearly everyone and every problem,

- Michél Hennig from moodstyler for layout and corporate design as well as operating www.hielc.org,
- our speakers, discussants, and the audience from all over the world for joining the conference,
- Pia Brandt, Eva Maria Bredler, Imke Frisch, Philip Heimann, Sebastian von Massow and Alexander Stark for doing tremendous editorial work to make this volume possible,
- Marie Sheldon and John Bennett at Brill Publishing for their commitment to working with us to present this volume to a worldwide scientific community,
- and last but not least the members of the board of the Internationaler Umweltrechtstag Hamburg e.V.

A "sea change" – did it already happen? Is it in a continuous process of happen? Will it ever happen? We invite you to study this volume – not to get final answers but to find inspiring ideas on some of the most topical issues of sustainable ocean governance.

Markus Kotzur, Nele Matz-Lück and Alexander Proelss
(Scientific Advisory Board of HIELC 2016)

List of Contributors

Douglas R. Burnett
is international Law Advisor for the International Cable Protection Committee (ICPC) (1999–2017), and Maritime Partner in the New York office of Squire Patton Boggs (U.S.) LLP.

Edwin Egede
is Senior Lecturer in International Law and International Relations at Cardiff University School of Law & Politics.

Pablo Ferrara
is Consultant at the Argentine National Congress and Of Counsel at Estudio O'Farrell (Buenos Aires) and Wöss and Partners (Mexico).

David Freestone
is Visiting Scholar and Professorial Lecturer at the George Washington University Law School in Washington D.C. He is Executive Director of the Sargasso Sea Commission.

Gabriele Goettsche-Wanli
has worked in the field of oceans at the United Nations for over 27 years, including as Director of the Division for Ocean Affairs and the Law of the Sea, Office of Legal Affairs (OLA), since 2013. She also served as Chief of the Treaty Section, OLA, for three years.

Vladimir V. Golitsyn
is President of the International Tribunal for the Law of the Sea.

Henning Jessen
is Associate Professor for Maritime Law & Policy at World Maritime University (WMU), Malmö.

Markus Kotzur
is Professor of European and International Law at the University of Hamburg and Director of Studies at the Europa-Kolleg Hamburg.

David Kenneth Leary
is Associate Professor, Faculty of Law, University of Technology Sydney.

Nele Matz-Lück
is Professor of Public Law and International Law, especially Law of the Sea at Kiel University and Co-director of the Walther Schücking Institute for International Law.

Alexander Proelss
is Professor of International and European Union Law at Trier University.

Joachim Sanden†
(1965–2017) was honorary Professor at the Leuphana University of Lüneburg and head of the soil protection department of the Hamburg Environmental Protection Authority (BUE).

Yoshifumi Tanaka
is Professor of International Law, Faculty of Law, University of Copenhagen.

Roda Verheyen
is partner at the law firm Günther in Hamburg, focusing on national and international environmental and planning law, and founder and former director of the Climate Justice Programme.

Rüdiger Wolfrum
is Professor Emeritus of international Law and Judge at the International Tribunal for the Law of the Sea (1996–2017). He is also managing director of the Max Planck Foundation for International Peace and the Rule of Law.

Keyuan Zou
is Harris Professor of International Law, University of Central Lancashire, United Kingdom.

PART 1

Sustainable Ocean Resource Governance – The Wider Horizons

CHAPTER 1

Opening Address

Vladimir V. Golitsyn

1 Introduction

Representatives of the Internationaler Umweltrechtstag Hamburg e.V,
Distinguished keynote speakers, panellists, and participants,
Ladies and gentlemen,

Welcome to the third Hamburg International Environmental Law Conference. It is a true honour for me to deliver this opening address, as well as to participate in today's activities. I am delighted at the sight of such a gathering of notable experts amongst us, here in the city of Hamburg. I wish to thank our host, the "Internationaler Umweltrechtstag Hamburg e.V." (IURT), its managing committee, and the scientific advisory board of the Hamburg International Environmental Law Conference, for putting together an exciting and timely programme that will allow us to reflect jointly on the essential matter of sustainable use of marine resources.

The title of this conference, 'A Sea Change for Sustainable Ocean Resource Governance', implies a shift in the international community's treatment of marine resources. Indeed, until as late as the mid-twentieth century, ocean resources were subject to a *laissez-faire* approach and treated as limitless, with little concern for their durability or for the environmental impact resulting from their exploitation. However, the United Nations Convention on the Law of the Sea ('the Convention'), and subsequent instruments, attest the undeniable emergence of a consensus that the exploitation of the oceans' riches resources requires sound management and conservation.

In order to contribute to sustainable use of ocean resources, one of the stated goals of this conference is to 'make recommendations for the establishment of new guidelines governing the sustainable use of ocean resources and mitigating threats to marine environments'. This is in line with recent developments at the international stage in relation to sustainable development, namely the global policy framework that has been adopted in September 2015 by the United Nations General Assembly as the Sustainable Development Goals or SDGs. In relation to oceans, SDG 14, which aims to 'conserve and sustainably use the oceans, seas and marine resources for sustainable development', draws

 | DOI 10.1163/9789004360273_002

attention to the role oceans can play in ensuring socio-economic wellbeing and intergenerational equity, and highlights the need for decoupling development on the one hand and marine environmental degradation on the other.

2 Addressing Suggested Topics

Allow me to make a few observations on the topics of today's keynote speeches, namely first, sustainable energy generation from the oceans, and second, sustainable management of ocean ecosystems. As I hope to participate in the working group on deep seabed mining, I will provide a few comments on the subject as well.

2.1 *Sustainable Energy Generation from the Oceans*

The world is hungrier than ever for energy. Increases in demographics and industrial activity have inevitably led to growing consumption of energy and energy-dependant products, thereby threatening the ecological balance on land, at sea, and in the atmosphere. This has prompted developments in the fields of science and technology, in particular as regards renewable energy from oceans. Marine renewable energy has attracted global private investment of over 200 billion US Dollars. But the benefits of access to such energy have not been uniform across the globe. In this regard, the transition of developing States from reliance on fossil fuel imports to designing alternative energy sources would greatly contribute to their development, while ensuring the long-term preservation of their environments.

In relation to the relevant legal framework, it would be valuable to examine the applicability of the Convention to marine renewable energy, and to identify new regulations and measures in place – in need of adoption – for governance of marine renewable energy generation and the protection of the marine environment. The examination of potential mechanisms for the transfer of marine technology would also be relevant in this respect. I am certain that Ms. Goettsche-Wanli's keynote speech on sustainable marine energy will enlighten our discussion and the exchanges of the working group on 'marine energy generation'.

2.2 *Sustainable Management of Ocean Ecosystems*

Turning to the next topic, the challenges faced by ocean ecosystems demonstrate an urgent need for their sustainable management, for conservation purposes but also for economic development and food security. The numbers speak for themselves. According to the United Nations, over three billion people depend on coastal and marine biodiversity.[1] Ocean ecosystems enjoy a global market value estimated at about 5 % of the global GDP (or 3 trillion US Dollars a year). Marine fisheries alone are responsible for the direct or indirect employment of over 200 million people worldwide. And yet, destructive fishing practices, illegal, unreported, and unregulated (IUU) fishing, the introduction of stressors to ecosystems from land, at sea, or from the atmosphere and loss of habitat, to name but a few, heavily affect as much as 40 % of the world's oceans.

Sturdier enforcement of international rules and regulations under the Convention and related instruments is therefore essential if we wish to continue to enjoy ecosystem services provided in the context of our global 'blue economy' responsibly. I wish to note in this respect the important contribution of the International Tribunal for the Law of the Sea to the application, interpretation, and development of international rules aimed at protecting the marine environment and its resources. Of particular relevance is the Tribunal's Advisory Opinion of 2 April 2015, delivered upon the request of the Sub-Regional Fisheries Commission, *inter alia* on the following matters:

- First, the obligations of flag States in cases where IUU fishing activities are conducted by vessels flying their flags within the Exclusive Economic Zone of third party States;
- Second, the extent to which flag States shall be held liable for IUU fishing activities conducted by vessels sailing under their flags; and
- Third, the respective liability of an international organisation and a flag State for the violation of the fisheries legislation of coastal States by a vessel to which a fishing license is issued within the framework of an international agreement concluded with the said organisation within its exclusive competence.

As to potential gaps in the governance system in the sustainable management of ecosystems and ways ahead, I look forward to hearing the intervention by Judge Wolfrum, and the ensuing discussion on the significant decision by the United Nations General Assembly to develop an international agreement on the sustainable use and conservation of biodiversity in areas beyond national jurisdiction under the Convention.

1 Cf. <http://www.un.org/sustainabledevelopment/oceans/> accessed June 2017.

2.3 *Deep Seabed Mining*

I now turn to deep seabed mining, the topic of the working group in which I will be taking part. As you know, the International Seabed Authority, the organization through which States Parties to the Convention 'organize and control activities in the Area',[2] has, as at 31 July 2015, approved 27 plans of work for exploration for marine mineral resources in the Area and had entered into 15-year contracts for exploration for polymetallic nodules, polymetallic sulphides and for cobalt-rich crusts.[3] Exploitation is also on the horizon. The Council of the Authority is expected to be presented in July of this year with a 'zero draft of the regulation for exploitation, including standard contract terms'.[4]

In light of the substantial current and expected activity in the Area, the Authority has an essential role to play as the body entrusted with the protection of the marine environment in the Area.[5] The Authority is involved in the 'elaboration and standardization ... of rules, regulations and procedures to ensure the effective protection of the marine environment'.[6] One of those initiatives is the environmental management plan for the Clarion-Clipperton Zone, approved in 2012, which includes the designation, on a provisional basis, of a network of areas of particular environmental interest. This plan was intended to be implemented over an initial three-year period with the view to undergo improvement in light of technological and scientific developments. The review of this management plan is expected to take place this year.[7]

Aside from the Authority, the responsibility of States should also be borne in mind. I wish to highlight in this respect the important contribution of the Seabed Disputes Chamber of the Tribunal, which, upon the request of the Council of the Authority, delivered its Advisory Opinion on 1 February 2011 on the 'Responsibilities and obligations of States with respect to activities in the Area'.[8] You may well recall that, in its Advisory Opinion, the Seabed Disputes Chamber addressed three questions concerning the following:

---

2 UNCLOS, Art. 157 (1).

3 UNGA Res 70/235 (23 December 2015), para. 54.

4 International Seabed Authority, 'Ongoing Development of Regulations on Exploitation of Mineral Resources in the Area', <https://www.isa.org.jm/legal-instruments/ongoing-development-regulations-exploitation-mineral-resources-area> accessed June 2017.

5 UNCLOS, Art. 145.

6 UNGA Res 70/235 (23 December 2015), para. 53.

7 Statement by Mr Nii Allotey Odunton, UNGA 70th session (8 December 2015), p. 3.

8 Responsibilities and Obligations of States Sponsoring Persons and Entities with Respect to Activities in the Area, Advisory Opinion, ITLOS Reports 2011, p.10.

- First, the legal responsibilities and obligations of States Parties to the Convention with respect to the sponsorship of activities in the Area in accordance with the Convention and the 1994 Agreement relating to the Implementation of Part XI of the Convention;
- Second, the extent of liability of a State Party for any failure to comply with the provisions of the Convention and the 1994 Agreement, by an entity whom it has sponsored; and
- Third, the necessary and appropriate measures that a sponsoring State must take in order to fulfil its responsibility under the Convention and the 1994 Agreement.

I hope that the Seabed Disputes Chamber's Advisory Opinion will offer guidance to the deliberations of the working group on seabed mining.

3 Closing

With this brief overview, I hope to have set the grounds for the work that lies ahead of us.

Lastly, allow me to express my gratitude to the city of Hamburg, which has exhibited leadership in its commitment as a 'smart city' to environmental protection. The city has been an indispensable partner of the Tribunal as its permanent seat since its inauguration in 1996, and will also be the host of the Tribunal's twentieth anniversary celebration scheduled for the fall of this year.

I very much look forward to hearing your views on the topics to be addressed in the working groups on deep seabed mining, seabed pipelines and cable systems, and marine energy generation, and to the elaboration of the recommendations.

I wish you all a successful and fruitful conference.

CHAPTER 2

Sustainable Production of Offshore Renewable Energy: A Global Perspective

Gabriele Goettsche-Wanli

1 Introduction

The United Nations Decade of Sustainable Energy for All,[1] the 2030 Agenda for Sustainable Development,[2] and the imperative in the Paris Agreement[3] to decarbonize energy provide a historic opportunity for the international community to strengthen its collective approach to achieving sustainable energy. The Goals in the 2030 Sustainable Development Agenda, and in particular Sustainable Development Goal 7, specifically call for States to 'enhance international cooperation to facilitate access to clean energy research and technology, including renewable energy, energy efficiency and advanced and cleaner fossil-fuel technology, and promote investment in energy infrastructure and clean energy technology'.[4] In 'The Future We Want,'[5] governments were urged *inter alia* to create enabling environments that facilitate public and private sector investment in relevant and needed cleaner energy technologies. The

* The opinions expressed in this article are personal, remain the responsibility of the author in her individual capacity and do not necessarily represent the views of the United Nations. The article builds on the keynote speech on 'Sustainable Energy Generation from and in the Oceans' which was presented by the author on behalf of the Division for Ocean Affairs and the Law of the Sea, Office of Legal Affairs, United Nations, with a powerpoint at the 3rd Hamburg International Environmental Law Conference, 15 and 16 April 2016. The following staff members in the Division provided valuable input to that powerpoint: Mr. Carlos Ivan Fuentes, Mr. Shawn Stanley and Ms. Snježana Žarić.

1 United Nations General Assembly Resolution 67/215 'Promotion of new and renewable sources of energy' (20 March 2013).

2 UNGA Res. 70/1 (21 October 2015).

3 Certified True Copy, United Nations Treaty Collection, at <https://treaties.un.org/doc/Treaties/2016/02/20160215%2006-03%20PM/Ch_XXVII-7-d.pdf> accessed 7 September 2017.

4 UNGA Res. 70/1, n. 2, p. 2.

5 UNGA Res. 66/288 (11 September 2012), p. 127.

 | DOI 10.1163/9789004360273_003

Secretary-General's initiative on 'Sustainable Energy for All'[6] supports these commitments by focussing on access to energy, energy efficiency and renewable energies.

The role of oceans as a potential source of renewable energy has been addressed by the United Nations General Assembly in its annual resolutions on oceans and the law of the sea,[7] and by its subsidiary body, the United Nations Open-ended Informal Consultative Process on Oceans and the Law of the Sea, at its thirteenth meeting in 2012.[8]

At the meeting of the Informal Consultative Process, delegations recognized that marine renewable energies offered unprecedented opportunities, but also presented environmental, economic and social challenges, especially to developing countries, including with regard to undertaking scientific research and acquiring technological knowledge.[9] The importance of government policies and financial support in encouraging investments in marine renewable energies was underscored. Particular emphasis was given to developing regulatory frameworks in conformity with international law. In this regard, the 1982 United Nations Convention on the Law of the Sea[10] (hereafter UNCLOS) was highlighted as the legal framework for any discussion on the development and production of marine renewable energies.[11]

---

6 Available at< www.se4all.org/sites/default/files/l/2013/09/SG_Sustainable_Energy_for_All_vision_final_clean.pdf> accessed 7 September 2017.

7 The General Assembly has welcomed the increasing attention being focused on oceans as a potential source of renewable energy and noted in this regard the summary of discussions of the United Nations Open-ended Informal Consultative Process on Oceans and the Law of the Sea at its thirteenth meeting on the topic of marine renewable energies. UNGA Res. 71/257 (20 February 2017), para. 278.

8 The meeting of the Informal Consultative Process provided an opportunity to consider the following: (1) various types of marine renewable energy technologies; their current uses and their role in sustainable development, including economic development, social development and environmental protection; (2) ongoing or planned marine renewable energies projects/work at the global and regional levels; and (3) opportunities and challenges related to marine renewable energies, including for cooperation and coordination. See UNGA 'Report on the work of the United Nations Open-ended Informal Consultative Process on Oceans and the Law of the Sea at its thirteenth meeting' (2 July 2012) UN Doc A/67/120.

9 UNGA 'Report on the work of the United Nations Open-ended Informal Consultative Process on Oceans and the Law of the Sea at its thirteenth meeting', n. 8, para. 49.

10 United Nations Convention on the Law of the Sea (adopted 10 December 1982, entered into force 16 November 1994) 1833 *UNTS* 3 (UNCLOS).

11 UNGA 'Report on the work of the United Nations Open-ended Informal Consultative Process on Oceans and the Law of the Sea at its thirteenth meeting', n. 8, para. 53.

UNCLOS sets out the legal framework within which all activities in the oceans and seas must be carried out, as reaffirmed annually by the United Nations General Assembly,[12] including for the production of renewable energy. As at 31 July 2017, there are 168 States parties to the Convention, including the European Union. In addition, many of its provisions are recognized as constituting customary international law.

Integration of the economic, social and environmental dimensions of sustainable development is at the core of the Convention. It creates an enabling legal environment for economic activities, such as the production of offshore renewable energy, including by providing for universally agreed limits for maritime zones and the peaceful settlement of disputes, as well as facilitating safety of navigation, the laying of submarine cables and pipelines and the operation of artificial islands, installations and structures. Indeed, the economic and social advancement of all peoples of the world is one of the objectives of UNCLOS. At the same time, the rights of States are balanced with duties to protect and preserve the marine environment and to ensure decent labour conditions. They are also balanced with the rights, duties and interests of other States.

This article will examine whether the current legal regime at the global level relating to offshore renewable energy provides the necessary legal certainty for technological development and investment, ensures the sustainable production of renewable energy from and in the oceans and also enables the accommodation of the various uses in the maritime zones in a balanced manner. While there are many facets of international law that need to be considered and integrated, this article will only focus on the law of the sea aspects. In that regard, the comprehensive legal framework provided by UNCLOS will be examined and how, complemented by other international instruments and integrated policy and management approaches and tools, such as marine spatial planning, it can provide a unifying legal and policy framework that can be further built upon, and an enabling legal environment for economic growth and social development while ensuring the protection and preservation of the marine environment. Particular attention will be given to the legal regime for offshore renewable wind energy, as it is currently the most developed form of offshore renewable energy.

12 See for example UNGA Res. 71/257, n. 7, preambular para. 5.

2 Types of Offshore Renewable Energy

Offshore energy includes not only conventional sources, such as oil, natural gas and coal, but also renewable energy. Renewables have been defined by the Intergovernmental Panel on Climate Change (IPCC) as any form of energy from solar, geophysical or biological sources that is replenished by natural processes at a rate that equals or exceeds its rate of use.[13] According to the IPCC, the potential of technically exploitable marine renewable energy, excluding offshore wind energy, is estimated at 7,400 exajoules per annum.[14] Estimates for near-shore offshore wind energy generation range from 15 to 130 exajoules per annum with a greater potential expected from deeper waters.[15] The installed capacity of offshore renewable energy, however, is unlikely to become significant until after 2020.[16] Offshore renewable energy, in particular, can provide the possibility of meeting an increasing global energy demand, while significantly contributing to carbon emission reductions, as well as having the capacity to deliver multiple economic, environmental and social benefits especially for small island developing States.[17] Renewable energy projects can be carried out as "Clean Development Mechanism" activities under the Kyoto Protocol to the United Nations Framework Convention on Climate Change.[18]

The following types of renewable energy currently are or can be produced from or in the oceans: 1) wind energy from turbines located in offshore areas; 2) kinetic energy derived from moving water masses, including tides, waves and currents; 3) ocean thermal energy conversion driven by temperature differences between the upper layers of seawater and colder deeper waters; 4) salinity gradient energy derived from the mixing of freshwater and seawater; 5) bioenergy derived from marine biomass; and 6) energy derived from floating solar power plants.[19]

---

13 Intergovernmental Panel on Climate Change (IPCC), *Special Report on Renewable Energy Sources and Climate Change Mitigation*, 2011, p. 164.

14 Ibid., p. 501.

15 Ibid., p. 539.

16 Ibid., p. 87.

17 Economic and Social Council 'Achieving sustainable development: integrating the social, economic and environmental dimensions. Concept note prepared for the Economic and Social Council integration meeting, held on 13 May 2013.' Available at <http://www.un.org/en/ecosoc/we/pdf/concept_note_2013.pdf> accessed 7 September 2017.

18 Kyoto Protocol to the United Nations Framework Convention on Climate Change, 1997, 2303 UNTS 162.

19 IPCC, n. 13; UNGA 'Report of the Secretary-General. Oceans and the law of the sea', UN Doc. A/67/79 (4 April 2012), made available to the thirteenth meeting of the UN Open-ended

2.1 *Wind Energy*

Offshore wind power is the most developed form of offshore renewable energy in terms of technology development, policy frameworks and installed capacity.[20] Offshore wind energy is harnessed from the kinetic energy of moving air. It is less mature than onshore wind energy extraction, is still developing and has opportunities for continued advancements. Offshore turbines are usually larger than onshore installations but are otherwise functionally similar in design. Turbine design and other project elements for offshore wind have benefitted significantly from research on and experience with land-based wind energy projects and offshore oil and gas development.[21]

Offshore wind farms have higher capital costs than onshore wind farms, but this can be offset to some extent by higher capacity factors. Investment in offshore wind in particular has been growing and as the technology has improved and demand has risen, costs have fallen. The world's exploitable offshore wind resources have been estimated at around an average of 22 terawatt-hours.[22] Recent statistics show energy capacities have expanded six-fold over the last six years and have expanded significantly in Asia, particularly China, and also throughout Europe, particularly in Belgium, Denmark, Germany, the Netherlands and the United Kingdom, with many countries introducing offshore wind energy.[23]

Informal Consultative Process on Oceans and the Law of the Sea when it discussed marine renewable energies in 2012; and P.E. Araghi et al., 'Other Marine-Based Energy Industries' in Group of Experts of the Regular Process, L. Inniss and A. Simcock, Joint Coordinators *The First Global Integrated Marine Assessment: World Ocean Assessment I.* United Nations Regular Process for Global Reporting and Assessment of the State of the Marine Environment, including Socioeconomic Aspects (United Nations, New York, and Cambridge University Press, United Kingdom and New York, 2017) [hereafter cited as '*The First Global Integrated Marine Assessment*']. The First Global Integrated Marine Assessment, which represents the outcome of the first cycle of the Regular Process for Global Reporting and Assessment of the State of the Marine Environment, including Socioeconomic Aspects, was welcomed with appreciation and its summary was approved by the General Assembly on 23 December 2015, in its resolution 70/235 on 'Oceans and the law of the sea' (15 March 2016), para. 266.

20 *The First Global Integrated Marine Assessment*, n. 19, p. 354.

21 Ibid.

22 Ibid., p. 355.

23 International Renewable Energy Agency (IRENA), *Renewable energy capacity statistics*, 2017, p. 25.

2.2 *Ocean Wave Energy, Tidal Power and Ocean Current Energy*

For other kinds of offshore renewable energy, including tide and wave energy, capacities have doubled over the last six years and have expanded significantly in Asia, particularly in the Republic of Korea, with many countries exploring various marine renewable possibilities.[24]

Ocean waves are generated by the transfer of energy from wind blowing over water. The global exploitable wave energy resource is estimated at around 3,700 GWa.[25] Floating or fixed conversion devices can be used to harness wave energy.

The global tidal energy capacity is estimated at 3,000 GWa, however, less than 3 per cent of this energy is located in areas suitable for power generation.[26] Energy derived from tides is most feasible in sites with large tidal ranges using barrages or submerged tidal turbines. The Sihwa Lake Tidal Power Station in the Republic of Korea, which has been operational since August 2011, is currently the world's largest tidal power barrage with a capacity of 254 MW.[27]

Ocean currents occur in the open ocean and are driven by winds and, on a global level, by the rotation of the Earth and the related natural physical forces acting on bodies of water. Energy from ocean currents, in particular those that are large and powerful, represent an enormous source of untapped energy which can be harnessed through underwater turbines, but the latter are currently still under development.[28]

2.3 *Ocean Thermal Energy Conversion*

Ocean thermal energy conversion (OTEC) is driven by the temperature differences which occur between the upper layers of seawater heated by the sun, and colder, deeper waters existing side by side. The required temperature differences of about 20°C (68°F) occur in tropical regions and could thus be a promising source of energy especially for small island developing States in the Caribbean and Indian and Pacific oceans.[29] Trials of onshore small-scale OTEC systems, however, continue to experience engineering challenges related to

24 Ibid., p. 15.

25 *The First Global Integrated Marine Assessment*, n. 19, p. 356.

26 Ibid.

27 Ibid.

28 Ibid., pp. 354 and 356.

29 A. Binger, 'Potential and Future Prospects for Ocean Thermal Energy Conversion (OTEC) In Small Island Developing States (SIDS)' (2004), available at <https://tethys.pnnl.gov/sites/default/files/publications/Binger-2004.pdf> accessed 7 September 2017.

pumping, vacuum retention and piping. In addition to existing onshore facilities, floating offshore facilities are also being developed.[30]

2.4 *Salinity Gradient Energy*

A salinity gradient is present where the mixing of freshwater and seawater occurs, for example at river mouths; this mixing releases energy as heat. Salinity gradient energy is harnessed either through the reversed electro-dialysis process which works on the difference in chemical potential between fresh and salt water, or the osmotic power process which is driven by the natural mixing tendency of fresh and salt water.[31]

While membrane technologies have advanced, they remain the main technical barrier to economical osmotic energy production.[32] Also, water on both sides must be low in particulates and other solids, eliminating many rivers from being a potential freshwater source.[33]

The first prototype osmotic power device became operational in 2009 in Norway but was discontinued in 2013 due to lack of investment.[34] The technology is currently experimental and still has technical challenges, as well as economic barriers to overcome.[35]

2.5 *Bioenergy*

Bioenergy is produced from biomass, such as seaweed or marine algae, through a variety of processes.[36] Algae can be harvested and processed into biofuels, including biodiesel and bioethanol.[37] Algae grown for biofuels can also provide a sink for carbon dioxide.[38] Microalgae are much more productive than agricultural crops.

Currently, most algae production takes place onshore, although offshore production is under development.[39] The National Aeronautics and Space Administration of the United States of America has been investigating the

---

30 For further information, see *The First Global Integrated Marine Assessment*, n. 19, p. 357.

31 IPCC, n. 13, p. 90.

32 J. Appiott, A. Dhanju, and B. Cicin-Sain, 'Encouraging renewable energy in the offshore environment', *Ocean and Coastal Management*, vol. 90, 2014, pp. 58–64.

33 *The First Global Integrated Marine Assessment*, n. 19, p. 355.

34 IRENA, *Salinity Gradient Energy Technology Brief*, 2014, p. 9.

35 Ibid., pp. 4–5.

36 *The First Global Integrated Marine Assessment*, n. 19, p. 355.

37 Ibid.

38 Ibid.

39 Ibid.

feasibility of growing algae in floating photobioreactors.[40] The Offshore Membrane Enclosure for Growing Algae (OMEGA)[41] technology aims at re-using the wastewater of coastal cities that is currently piped out and released into the ocean. Fuelled by the sun and carbon dioxide from the atmosphere, the algae would eat the waste and produce oils that can be converted to fuel.

2.6 *Solar Energy*

Offshore solar energy uses the energy from the sun's radiation through floating arrays offshore. While floating solar arrays already exist inland on bodies of water, this technology is still under development for offshore areas, although an increasing number of States are turning to it. Compared to mounted panels, floating systems are naturally cooled by the bodies of water they sit on, therefore boosting power production efficiency.[42] Japan opened its first offshore floating solar power plant in November 2013.[43] The world's largest floating solar power plant is located in the city of Huainan, in China and has a capacity of 40 MW.[44] Other floating solar projects are under development in Australia, Brazil, Japan, the United Kingdom and the United States of America.[45]

3 States' Rights to Produce Offshore Renewable Energy in the Various Maritime Zones

A global treaty which is of central importance for the regulation of activities relating to offshore renewable energy is UNCLOS. This section will begin with a description of the rights and obligations of States in the various maritime zones (internal waters, archipelagic waters, the territorial sea, exclusive

40 Ibid., p. 357.

41 Described in detail at <https://www.nasa.gov/centers/ames/research/OMEGA/#.V93hAfArK-W> accessed 7 September 2017.

42 L. Chow, '5 Floating Solar Farms Helping Power the World' (24 May 2016) available at <http://www.ecowatch.com/5-floating-solar-farms-helping-power-the-world-1891141371.html> accessed 7 September 2017.

43 L. Zimmer, 'Kyocera Opens Japan's Largest Offshore Solar Power Plant (11 May 2013) available at <http://inhabitat.com/kyocera-opens-japans-largest-offshore-solar-power-plant> accessed 7 September 2017.

44 S. Brandon, 'China just switched on the world's largest floating solar power plant' (2 June 2017) available at <https://www.weforum.org/agenda/2017/06/china-worlds-largest-floating-solar-power> accessed 7 September 2017.

45 L. Chow, n. 42.

economic zone (EEZ), continental shelf, the high seas and the Area) in relation to the production of offshore renewable energy. Subsequent sections will focus on the environmental and social aspects, followed by an examination of the legal regime governing the technology used for the production of offshore renewable energy. The legal regime to address security threats to the production of offshore renewable energy, as well as requirements regarding compliance, enforcement, responsibility and liability and settlement of disputes will also be addressed. The article will conclude with a brief examination of the available tools for strengthening cooperation and coordination at all levels. Since UNCLOS is a framework Convention, which relies for its effective implementation on the further development of international instruments, the Convention's provisions that have been further developed by other legal instruments will also be presented. Given its focus on the global level, this article will only provide some illustrative examples of relevant developments at the regional level.

UNCLOS defines the extent of States rights and duties in the different maritime zones. A specific reference to the production of energy from water, currents and winds is only made in Part V of UNCLOS addressing the legal regime for the EEZ.

In internal waters and archipelagic waters and in the territorial sea – which can extend to a maximum breadth of 12 nautical miles measured from the baselines, and include the airspace over the territorial sea as well as its bed and subsoil – coastal and archipelagic States exercise sovereignty.[46] Although not specifically mentioned, the exercise of sovereignty by coastal and archipelagic States includes the right to produce renewable energy. Other States can exercise the right of innocent passage in internal waters enclosed by straight baselines which had not previously been considered as such, through archipelagic waters and through the territorial sea.[47] Rights of transit passage and archipelagic sea lanes passage can be exercised in straits used for international navigation and archipelagic sea lanes.[48]

UNCLOS provides that the EEZ can extend to a maximum distance of 200 nautical miles from the baselines from which the breadth of the territorial sea is measured.[49] In that zone, the coastal State has sovereign rights for the purpose of exploring and exploiting, conserving and managing the natural resources, whether living or non-living, of the water superjacent to the seabed and of the

---

46 UNCLOS, n. 10, see generally Parts II and IV.

47 Ibid., Article 8, Part II, section 3, and Article 52.

48 Ibid., see generally Part III and Articles 53 and 54.

49 Ibid., Article 57.

seabed and its subsoil, and with regard to other activities for the economic exploration and exploitation of the zone, such as the production of energy from water, currents and winds.[50] This provision is not exhaustive and can extend to other types of offshore renewable energy, such as tidal power, OTEC and bioenergy.[51] UNCLOS requires that the rights with respect to the seabed and subsoil be exercised in accordance with Part VI of the Convention on the continental shelf.[52] In the EEZ, the coastal State has jurisdiction with regard to, *inter alia*, the protection and preservation of the marine environment and the establishment and use of artificial islands, installations and structures.[53] Under Article 60 of UNCLOS, the coastal State has exclusive jurisdiction over such artificial islands, installations and structures, as explained in more detail below.

In exercising its rights and performing its duties under the Convention, the coastal State must have due regard to the rights and duties of other States and act in a manner compatible with the provisions of UNCLOS.[54] All other States can exercise freedom of navigation and overflight in the EEZ, as well as freedom to lay submarine cables and pipelines, and other internationally lawful uses of the sea related to these freedoms, such as those associated with the operation of ships and submarine cables and pipelines, and compatible with the other provisions of the Convention.[55] In exercising their rights and performing their duties under the Convention in the EEZ, States must have due regard to the rights and duties of the coastal State and comply with its laws and regulations adopted in accordance with the Convention and other rules of international law insofar as they are not incompatible with Part V.[56]

Part VI of UNCLOS grants the coastal State sovereign rights over the continental shelf for the purpose of exploring and exploiting the mineral and other non-living resources of the seabed and subsoil, as well as living resources belonging to sedentary species.[57] The continental shelf can extend beyond

---

50 Ibid., Article 56, para. 1(a).

51 M.A. Castelos, 'Marine Renewable Energies: Opportunities, Law and Management', *Ocean Development and International Law*, vol. 45(2), 2014, p. 226; S. Mc Donald and D.L. VanderZwaag, 'Renewable Ocean Energy and the International Law and Policy Seascape: Global Currents, Regional Surges', in A. Chircop et al., *Ocean Yearbook*, vol. 29, Brill 2015, p. 303.

52 UNCOS, n. 10, Article 56, para. 3.

53 Ibid., Article 56, para. 1(b).

54 Ibid., Article 56, para. 2.

55 Ibid., Article 58, paras. 1 and 2.

56 Ibid., Article 58, para. 3.

57 Ibid., Article 77.

200 nautical miles from the baselines from which the breadth of the territorial sea is measured up to its outer limits as defined under Article 76. The rights of the coastal State over the continental shelf do not depend on occupation, effective or notional, or on any express proclamation.[58] However, unlike Article 56, which makes specific reference to the production of energy from water, currents and wind, Article 77 of UNCLOS only mentions the sovereign rights of the coastal State over the continental shelf for the purpose of exploring and exploiting its natural resources. Thus, where a 200 nautical mile EEZ has been proclaimed by the coastal State,[59] the right of the coastal State to produce renewable energy on the continental shelf is clear. However, where the continental shelf extends beyond 200 nautical miles from the baselines from which the breadth of the territorial sea is measured up to its outer limits as defined under Article 76, it is also necessary to consider the rights of States on the high seas, as well as the extent of the rights and duties of coastal States with respect to artificial islands, installations and structures and submarine cables and pipelines on the continental shelf. UNCLOS prescribes that the rights of the coastal State over the continental shelf do not affect the legal status of the superjacent waters or of the airspace above those waters[60] and the exercise of the rights of the coastal State over the continental shelf must not infringe or result in any unjustifiable interference with navigation and other rights and freedoms of other States as provided for in the Convention.[61]

There are divergent views regarding the extent of the coastal State's rights with respect to offshore renewable energy on the continental shelf beyond 200 nautical miles from the baselines from which the breadth of the territorial sea is measured depending on the interpretation that is given to Article 80 of UNCLOS,[62] which applies Article 60 *mutatis mutandis* to artificial islands, installations and structures on the continental shelf. Article 60 grants the coastal State the exclusive right to construct and to authorize and regulate the construction, operation and use of *inter alia* installations and structures for the purposes provided for in Article 56 and other economic purposes.[63]

58 Ibid., para. 3.

59 The rights of the coastal State with respect to the EEZ need to be established; they do not exist *ab initio* as in the case of the continental shelf. R.R. Churchill and A.V. Lowe, A.V., *The law of the sea*, Manchester University Press, 1999, p. 145.

60 UNCLOS, n. 10, Article 78, para. 1.

61 Ibid., Article 78, para. 2.

62 For a detailed analysis, see A. Chircop and P. L'Esperance, 'Functional Interactions and Maritime Regulation: The Mutual Accommodation of Offshore Wind Farms and International Navigation and Shipping', *Ocean Yearbook*, vol. 30, 2016, pp. 439–487.

63 UNCLOS, n. 10, Article 60, para. 1(b).

According to one view, Article 60 applies to Article 80 insofar as it can be applied to the continental shelf.[64] This interpretation would mean that the exclusive right of the coastal State to construct, authorize and regulate the construction, operation and use of any artificial island, installation or structure on its continental shelf beyond 200 nautical miles from the baselines from which the breadth of the territorial sea is measured and on the continental shelf within 200 nautical miles where a proclaimed EEZ does not extend to 200 nautical miles or where such zone has not been proclaimed would be limited to the exploration and exploitation of the mineral and other non-living resources of the seabed and subsoil. A contrary point of view holds that Article 80 considered together with Article 246, which requires *inter alia* the consent of the coastal State for marine scientific research on the continental shelf if the project involves the construction, operation or use of artificial islands, installations and structures referred to in Articles 60 and 80, means that UNCLOS grants each coastal State the exclusive right to construct, authorize and regulate the construction, operation and use of any artificial island, installation or structure on its continental shelf including for the production of energy.[65] The latter interpretation would also be supported by Article 79 (4) of UNCLOS which seems to differentiate between the exercise of jurisdiction by the coastal State over cables and pipelines constructed or used in connection with the exploration of the continental shelf or the exploitation of its resources, and the cables and pipelines used in the operations of artificial islands, installations and structures under its jurisdiction.

On the high seas, which comprise all parts of the sea that are not included in the EEZs, territorial seas, archipelagic waters or internal waters of States,[66] all States enjoy freedoms with regard to certain uses, including freedom of navigation and freedom to lay submarine cables and pipelines and to construct arti-

64 S.N. Nandan and S. Rosenne (eds.), *United Nations Convention on the Law of the Sea: A Commentary*, vol. II, Martinus Nijhoff, 1993, p. 926.

65 V.P. Nanda, 'The Legal Framework for the Development of Ocean Thermal Energy Conversion (OTEC)', *San Diego Law Review*, vol. 19, 1982, pp. 385–408; S. Joseph, 'Legal Issues Confronting the Exploitation of Renewable Sources of Energy from the Oceans', *California Western International Law Journal*, vol. 11, 1981, pp. 394–395; J.W. Kindt, 'Ocean Thermal Energy Conversion', *Georgia Journal of International and Comparative* Law, vol. 14 no. 1, 1984, p. 16; W. Nielsen and T. Davenport, 'Submarine Cables and Offshore Energy' in D. Burnett, R. Beckman, T. Davenport (eds), *Submarine Cables: The Handbook of Law and Policy*, Brill, 2014, p. 369; K. Scott, 'Tilting at offshore windmills: Regulating wind farm development within the renewable energy zone', *Journal of Environmental Law*, vol. 18 no. 1, 2006, p. 96.

66 UNCLOS, n. 10, Article 86.

ficial islands and other installations permitted under international law, subject to Part VI of the Convention.[67] No State may validly purport to subject any part of the high seas to its sovereignty.[68]

While the technology is not yet available to enable the development of renewable energy on the high seas, early consideration could nonetheless be given to how the right of all States to exercise freedom of the high seas could be reconciled with future potential permanent energy projects developed by the private sector.[69] During the thirteenth meeting of the UN Open-ended Informal Consultative Process on Oceans and the Law of the Sea, questions were raised over the ownership and the transmission of energy to land-based facilities from areas beyond national jurisdiction, as well as the appropriate forum for the resolution of possible jurisdictional issues.[70] Some panellists suggested that these issues should be addressed at the international level.[71] In that regard, it has been suggested that an international oversight body could encourage development of renewable energy projects while ensuring that any such project is in conformity with international law.[72]

UNCLOS provides in Part XI, that the Area, which comprises the seabed and ocean floor and subsoil thereof, beyond the limits of national jurisdiction and its resources are the common heritage of mankind.[73] For the purposes of Part XI, resources mean all solid, liquid or gaseous mineral resources *in situ* in the Area at or beneath the seabed, including polymetallic nodules.[74] Installations can be erected and placed to carry out activities in the Area subject to the rules, regulations and procedures of the International Seabed Authority.[75] UNCLOS defines "activities in the Area" to mean 'all activities of exploration for, and exploitation of, the resources of the Area'.[76] The Convention provides that other activities in the marine environment must be conducted with reasonable regard for activities in the Area, just as activities in the Area have to be carried out with reasonable regard for other activities in

67 Ibid., Article 87.

68 Ibid., Article 89.

69 N.J. Lund, 'Renewable Energy as a Catalyst for Changes to the High Seas Regime', *Ocean and Coastal Law* Journal, vol. 15, 2010, p. 101.

70 UNGA 'Report on the work of the United Nations Open-ended Informal Consultative Process on Oceans and the Law of the Sea at its thirteenth meeting, n. 8, para. 44.

71 Ibid.

72 N.J. Lund, n. 69, pp. 101, 124–125.

73 UNCLOS, n. 10, Article 1, para. 1, and Article 136.

74 Ibid., Article 133.

75 Ibid., Article 147.

76 Ibid., Article 1, para. 3.

the marine environment.[77] The Authority is mandated by UNCLOS *inter alia* to adopt appropriate rules, regulations and procedures for the protection of the marine environment[78] and for the protection of human life.[79]

During the Third United Nations Conference on the Law of the Sea proposals to expand the definition of resources in Part XI to include 'water, steam, hot water'[80] were not accepted.[81] Views remain divided regarding the meaning of "resources" in UNCLOS.[82] It has been remarked that energy harnessed through offshore renewable energy should be considered a natural resource contrary to past conceptions when such energy could not be exploited, as it is now possible to do so and is economically viable.[83] Article 56 of UNCLOS characterizes natural resources as either living or non-living.

In sum, UNCLOS can generally be considered as providing sufficient general guidance on the rights of States to produce offshore renewable energy in areas within national jurisdiction, except in the case of the continental shelf beyond 200 nautical miles from the baselines from which the breadth of the territorial sea is measured and in the case of the continental shelf within 200 nautical miles where a proclaimed EEZ does not extend to 200 nautical miles or where such zone has not been proclaimed. Clarifications regarding the applicable legal regime may also be required in the future with respect to the high seas, ideally well before technological advances enable the production of energy in that maritime zone. Of course, ultimately, legal certainty is also dependent on clear delimitation of maritime areas, so that the allocation of jurisdiction is equally clear.[84]

---

77 Ibid., Article 147.

78 Ibid., Article 145.

79 Ibid., Article 146.

80 'Informal Composite Negotiating Text/Revision 1' (28 April 1979) A/CONF.62/WP.10/Rev.1, p. 73.

81 In the Informal Composite Negotiating Text/Revision 2 (ICNT/Rev.2) 'resources' were defined as 'mineral resources *in situ*' and the words 'water, steam, and hot water' were deleted from the list. See (11 April 1980) A/CONF.62/WP.10/Rev.2, p. 75.

82 F. Galea, 'A Legal Regime for the Exploration and Exploitation of Offshore Renewable Energy', *Ocean Yearbook*, vol. 25, 2011, pp. 105–107; V.P. Nanda, n. 65, pp. 392–93; Joseph, n. 65, pp. 387, 393; and J.W. Kindt, n. 65, p. 18.

83 F. Galea, n. 82, pp. 106–107; F.H. Th. Wegelein, *Marine Scientific Research. The Operation and Status of Research Vessels and other Platforms in International Law*, Martinus Nijhoff, Leiden, 2005, p. 87.

84 M. Gavouneli, 'Energy Installations in the Marine Environment' in J. Barrett and R. Barnes (eds.), *Law of the Sea: UNCLOS as a living treaty*, British Institute of International and Comparative Law, 2016, pp. 187–208. The article is also available at <http://www

4 The Environmental Dimension

A move from traditional sources of energy to the use of offshore renewable energy will clearly benefit the environment. But at the same time, offshore renewable energy projects have the potential to adversely impact the marine environment as indicated also in the First Global Integrated Marine Assessment.[85] Potential adverse impacts include habitat loss, change or degradation; direct mortality; injurious noise and electromagnetic field impacts. Living marine resources may be affected by these impacts, which can occur during the construction, operation and decommissioning phases.[86] They can also affect birds, especially migratory birds and bats, which might collide with turbines.

The activities and effects associated with the construction and decommissioning of renewable energy source devices and installations (e.g. seismic explorations; construction noise caused by drilling, use of explosives, ramming and piling; dredging; cable laying; water turbidity; construction vessel activities) may have short- or medium-term impacts, whilst the electromagnetism and physical presence of structures could have long-term impacts.[87] It is therefore important to take the variation in duration of impacts into account in an environmental impact assessment of offshore renewable energy.[88]

At the thirteenth meeting of the UN Open-ended Informal Consultative Process on Oceans and the Law of the Sea, several delegations recalled the need to give due consideration to the potential adverse environmental, social and cultural impacts of marine renewable energies.[89] The need to further study the impacts of tidal barrages on the marine environment was specifically highlighted.[90] At its session in September 2016, the World Conservation Congress asked States and competent authorities *inter alia* to implement a strategy for the development of offshore renewable energy (including, for example, adequate funding mechanisms), which takes environmental issues

.energyatsea.law.uoa.gr/uploads/media/Gavouneli_BIICL.pdf> accessed on 7 September 2017, p. 3.

85 *The First Global Integrated Marine Assessment*, n. 19, pp. 357–358.

86 Ibid.

87 UNGA 'Report of the Secretary-General. Oceans and the law of the sea', n. 19, p. 22.

88 S. Mc Donald and D.L. VanderZwaag, n. 51, p. 301.

89 UNGA 'Report on the work of the United Nations Open-ended Informal Consultative Process on Oceans and the Law of the Sea at its thirteenth meeting', n. 8, para. 58.

90 Ibid., para. 39.

into account through the entire life cycle of the project, and subject this strategy to rigorous strategic environmental assessment.[91]

UNCLOS sets out in its Part XII and in several other provisions a comprehensive framework for the protection and preservation of the marine environment. All States have the obligation to protect and preserve the marine environment.[92] They must *inter alia* take, individually or jointly as appropriate, all measures consistent with UNCLOS that are necessary to prevent, reduce and control pollution of the marine environment from any source, using for this purpose the best practicable means at their disposal and in accordance with their capabilities, and they must endeavour to harmonize their policies in this connection.[93] States must ensure that activities under their jurisdiction or control do not cause transboundary pollution.[94]

Pollution, as defined in Article 1 of UNCLOS, includes the introduction by man, directly or indirectly, of energy into the marine environment which results or is likely to result in such deleterious effects as harm to living resources and marine life, amongst others. The term "marine life" is also included in Part XII,[95] where also the terms "natural resources",[96] "ecosystems"[97] and "species"[98] are mentioned. UNCLOS specifically requires States in Article 194 (5) to take measures 'necessary to protect and preserve rare or fragile ecosystems as well as the habitat of depleted, threatened or endangered species and other forms of marine life'. It has been suggested that the definition of pollution in Article 1 is broad enough to include noise pollution.[99]

The general provisions in UNCLOS are supplemented by duties to cooperate on a global and, as appropriate, regional basis,[100] monitoring and environmental impact assessment,[101] notification and contingency plans,[102] the

91 IUCN Res WCC-2016-Rec-112 in IUCN, *IUCN Resolutions, Recommendations and other Decisions*, Gland, 2016, p. 243.

92 UNCLOS, n. 10, Article 192.

93 Ibid., Article 194, para. 1.

94 Ibid., Article 194, para. 2.

95 Ibid., Article 194, para. 5.

96 Ibid., Article 193.

97 Ibid., Article 194, para. 5.

98 Ibid.

99 E. McCarthy, *International Regulation of Underwater Sound: Establishing Rules and Standards to Address Ocean Noise Pollution*, Kluwer, 2004, p. 124; S. Mc Donald and D.L. VanderZwaag, n. 51, p. 304.

100 UNCLOS, no 10, Article 197.

101 Ibid., Articles 204–206.

102 Ibid., Articles 198 and 199.

adoption and enforcement of laws and regulations, State responsibility and liability[103] and the provision of scientific and technical assistance to developing States.[104]

The Convention specifically mandates the further development of legal instruments. In particular, it requires internationally accepted rules, standards and recommended practices and procedures to be applied as minimum standards in the development and enforcement of national laws, regulations and measures for pollution from activities in the Area,[105] from seabed activities subject to national jurisdiction,[106] from vessels,[107] and by dumping.[108] With regard to pollution from land-based sources[109] and pollution from or through the atmosphere, internationally accepted rules, standards and recommended practices and procedures must also be taken into account in the development of national laws and regulations.[110]

The legal regime in UNCLOS is thus complemented by a wide range of legal instruments, as also exemplified below. Customary international law further supplements this regime, for example in relation to the obligation to undertake an environmental impact assessment in regard to any proposed activity which may have a significant adverse impact in a transboundary context. In the *Pulp Mills* case,[111] the International Court of Justice specified that an environmental impact assessment 'must be conducted prior to the implementation of the project' and that 'once operations have started and, where necessary, throughout the life of the project, continuous monitoring of its effects on the environment shall be undertaken'.[112]

The application of environmental impact assessments, including strategic environmental assessments, and area-based management tools in areas beyond national jurisdiction have also been considered by the Preparatory Committee established by General Assembly resolution 69/292 Development of an international legally binding instrument under UNCLOS on the conservation and sustainable use of marine biological diversity of areas beyond

103 Ibid., Article 235.
104 Ibid., Articles 202 and 203.
105 Ibid., Articles 209 and 215.
106 Ibid., Articles 208 and 214.
107 Ibid., Articles 211, 217, 218, 219 and 220.
108 Ibid., Articles 210 and 216.
109 Ibid., Articles 207 and 213.
110 Ibid., Articles 212 and 222.
111 *Case concerning Pulp Mills on the River Uruguay* (Argentina v. Uruguay), Judgment, ICJ Reports 2010, p. 10.
112 Ibid., p. 83, para. 205.

national jurisdiction.[113] Following the completion of its fourth session in July 2017, the Preparatory Committee submitted its recommendations to the General Assembly, including elements to be considered with a view to the development of a legally binding instrument under UNCLOS.[114] The elements do not reflect consensus.[115] They include non-exclusive elements that generated convergence among most delegations.[116] Included among these are elements on environmental impact assessments and strategic environmental assessments. Whether the instrument should address strategic environmental assessments is identified specifically as an issue requiring further discussion.[117]

Guidance on the conduct of environmental impact assessments and strategic environmental assessments has been developed at the global level. The measures adopted under the Convention on Biological Diversity[118] (CBD) and the Convention on Wetlands of International Importance especially as Waterfowl Habitat (Ramsar Convention)[119] are particularly noteworthy for offshore renewable energy projects.

The Voluntary Guidelines for the Consideration of Biodiversity in Environmental Impact Assessments and Strategic Environmental Assessments in Marine and Coastal Areas were adopted by the Conference of the Parties to the CBD in 2012,[120] pursuant to Article 14 of the CBD which includes requirements regarding environmental impact assessment which apply to processes and activities, regardless of where their effects occur, carried out under its parties' jurisdiction or control. The Guidelines list noise inputs among the direct drivers of biophysical change with known impacts on biodiversity and associated ecosystem services.[121] They also recommend including noise emissions

113 UNGA 'Report of the Preparatory Committee established by General Assembly resolution 69/292: Development of an international legally binding instrument under the United Nations Convention on the Law of the Sea on the conservation and sustainable use of marine biological diversity of areas beyond national jurisdiction', UN Doc. A/AC.287/2017/PC.4/2 (31 July 2017).

114 Ibid., para. 38.

115 Ibid.

116 Ibid., section A.

117 Ibid., section B.

118 Convention on Biological Diversity, 1992, 1760 *UNTS* 79.

119 Convention on Wetlands of International Importance especially as Waterfowl Habitat, 1971, 996 *UNTS* 246, ['Ramsar Convention'].

120 CBD Decision XI/18 (5 December 2012), pp. 7–8; and (21 August 2012) UNEP/CBD/COP/11/23, annex (21 August 2012).

121 UNEP/CBD/COP/11/23, n. 120, annex, Box 3 at p. 37.

in the screening criteria for determining the need for and required level of environmental impact assessment in relation to areas providing key ecosystem services.[122]

The Guidance for addressing the implications for wetlands of policies, plans and activities in the energy sector, including tidal and wave energy projects, adopted by the Conference of the Parties to the Ramsar Convention in 2012,[123] also recommends *inter alia* the rigorous application of strategic environmental assessments and environmental impact assessment processes to proposed renewable energy projects that may alter the ecological character of wetlands and urges the adoption of a precautionary approach when energy sector activities may seriously or irreversibly impact Ramsar Wetlands of International Importance or other internationally important wetlands, or when the strategic environment assessment or environmental impact assessment predicts any substantial or irreversible loss of wetland ecosystem services.[124]

Among the regional instruments that have been developed,[125] the Convention on Access to Information, Public Participation in Decision-Making and Access to Justice in Environmental Matters[126] is particularly noteworthy. It sets out procedural requirements linked to the participatory rights of citizens, such as environmental impact assessments and access to justice.

122 Ibid., appendix 1, p. 23; S. Mc Donald and D.L. VanderZwaag, n. 51, p. 307.

123 Ramsar Convention, Res XI.10, available at <http://www.ramsar.org/sites/default/files/documents/pdf/cop11/res/cop11-res10-e.pdf> accessed 7 September 2017.

124 S. Mc Donald and D.L. VanderZwaag, n. 51, pp. 310–311.

125 See, for example, Convention on Environmental Impact Assessment in a Transboundary Context, 1991, 1989 UNTS 309; the Protocol on Strategic Environmental Assessment under the Convention on Environmental Impact Assessment in a Transboundary Context, 2003, 2685 UNTS 140; the Convention on Access to Information, Public Participation in Decision-Making and Access to Justice in Environmental Matters, 1998, 2161 UNTS 447; the Convention on the Protection of the Marine Environment of the Baltic Sea Area, 1974, 1507 UNTS 167; Convention for the Protection of the Marine Environment of the North-East Atlantic, 1992, 2354 UNTS 67, and the OSPAR Guidance on Environmental Considerations for Offshore Wind Farm Development, Reference number: 2008–3, available at <https://www.ospar.org/documents?d=32631> accessed 7 September 2017; Convention for the Protection of the Mediterranean Sea against Pollution, 1976, 1102 UNTS 27, amended in 1995 and renamed the Convention for the Protection of the Marine Environment and the Coastal Region of the Mediterranean; Protocol on Environmental Protection to the Antarctic Treaty, 2001, available at <https://www.ats.aq/documents/recatt/Att006_e.pdf> accessed 7 September 2017.

126 Convention on Access to Information, Public Participation in Decision-Making and Access to Justice in Environmental Matters, n. 125.

Furthermore, in response to the potential or actual impacts on biodiversity from renewable energy technology deployments,[127] a number of measures have been taken at the global and regional levels. At the global level, resolutions have been adopted by the Conference of the Parties to the Convention on the Conservation of Migratory Species of Wild Animals[128] (CMS) and Guidelines endorsed in 2014 on how to avoid or mitigate impacts on migratory species of the deployment of renewable energy technologies, including those used for the production of offshore renewable energy.[129] The Guidelines recommend carrying out an environmental impact assessment and strategic environmental assessment and *inter alia* set out best practice on mitigating and avoiding possible conflicts between offshore renewable energy technologies and migratory species. Such practice includes a site selection and review process to avoid locating the development in major migration corridors or sensitive habitats and avoiding scheduling construction, maintenance, and decommissioning activities during important migration periods when migratory species would potentially be in the area to reduce negative interactions with migratory wildlife. Recommended mitigation measures include minimizing the use of slack or loose tether and anchor lines to reduce entanglement risk to species; use of observers on board vessels to inform temporary cessation of construction, maintenance, and decommissioning activities with the aim of avoiding disturbance to marine species in the work area, including sea turtles and marine mammals; using noise-deflecting devices (e.g. bubble walls or baffles) around the work site during high-decibel generating phases of construction to avoid physiological impacts to marine mammals and sea turtles; and burying undersea cables within the ocean energy development array and at the landfall

127 F. Galea, n. 82, p. 125.

128 Convention on the Conservation of Migratory Species of Wild Animals, 1979, (CMS) 1651 UNTS 356. Relevant resolutions in relation to reconciling renewable energy development with migratory species conservation include: CMS Res 7.5 (24 September 2002), available at <http://www.cms.int/sites/default/files/document/RES_7_05_Wind_Turbine_0_0.pdf> accessed 7 September 2017; CMS Res 9.19 (5 December 2008), available at <http://www.ascobans.org/sites/default/files/document/AC16_46_CMSRes9-19_OceanNoise_1.pdf> accessed 7 September 2017; CMS Res 10.24 (25 November 2011), available at <http://www.cms.int/sites/default/files/document/10_24_underwater_noise_e_0_0.pdf> accessed 7 September 2017; CMS Res 11.27 (9 November 2014), available at <http://www.cms.int/raptors/sites/default/files/document/mos2_inf14_cms_res_11_27_e.pdf> accessed 7 September 2017.

129 CMS Res 11.27, n. 128; and 'Renewable Energy Technologies and Migratory Species: Guidelines for Sustainable Deployment' (2 October 2014) UNEP/CMS/COP11/Doc.23.4.3.2.

connection to depths within the sediment that will minimize or eliminate the impacts from electromagnetic fields to sea turtles and marine mammals.[130]

The International Whaling Commission has addressed the interactions between offshore renewable energy projects, in particular offshore wind farms, tidal-stream driven devices and wave energy converters, and cetaceans through its Scientific Committee which at a workshop in 2012 identified a number of mitigation and other measures that can be taken to address possible interactions, including the impact of noise, collisions and entanglement in tethering lines.[131]

Recognizing that the impact of underwater-radiated noise from commercial ships may have both short- and long-term negative consequences on marine life, especially marine mammals, the International Maritime Organization (IMO) approved non-mandatory Guidelines in order to provide general advice about reduction of underwater noise to designers, shipbuilders and ship operators.[132]

At the regional level, measures have been adopted within the framework of the following treaties: the Convention for the Protection of the Marine Environment of the North-East Atlantic,[133] the Convention for the Protection of the Marine Environment and the Coastal Region of the Mediterranean,[134] the Agreement on the Conservation of African-Eurasian Migratory Waterbirds[135] (AEWA), the Agreement on the Conservation of Populations of European Bats[136]

130 Ibid., pp. 51–52.

131 'Report of the IWC Scientific Committee Workshop on Interactions between Marine Renewable Projects and Cetaceans Worldwide', IWC SC/64/Rep6 Rev1.

132 International Maritime Organization (IMO), 'Guidelines for the reduction of underwater noise from commercial shipping to address adverse impacts on marine life' (7 April 2014), IMO Doc MEPC.1/Circ.833.

133 Convention for the Protection of the Marine Environment of the North-East Atlantic, n. 125; OSPAR Commission, Agreement 2008–3, 'Guidance on Environmental Considerations for Offshore Wind Farm Development'; OSPAR Commission, Agreement 2012–2 'Guidelines on Best Environmental Practice (BEP) in Cable Laying and Operation'.

134 Convention for the Protection of the Marine Environment and the Coastal Region of the Mediterranean, n. 125.

135 Agreement on the Conservation of African-Eurasian Migratory Waterbirds, 1996 2365 *UNTS* 203. See also AEWA Res 5.16 (18 May 2012), available at <http://www.cms.int/sites/default/files/document/ETF1_Inf.7.pdf> accessed 7 September 2017.

136 Agreement on the Conservation of Bats in Europe, 1991 (1994) UKTS No. 9. See also EUROBATS Res 7.5 (September 2004) EUROBATS.MoP7.Record.Annex8, available at <http://www.eurobats.org/sites/default/files/documents/pdf/Meeting_of_Parties/

(EUROBATS), the Agreement on the Conservation of Small Cetaceans in the Baltic, North East Atlantic, Irish and North Seas[137] (ASCOBANS), the Agreement on the Conservation of Cetaceans in the Black Sea, Mediterranean Sea and Contiguous Atlantic Area[138] (ACCOBAMS), and the 1979 Convention on the Conservation of European Wildlife and Natural Habitats (Bern Convention).[139]

A Multi-stakeholder Task Force on Reconciling Selected Energy Sector Developments with Migratory Species Conservation[140] was established in 2014 to assist Parties or Signatories to CMS, AEWA, EUROBATS, ASCOBANS, ACCOBAMS, the Memorandum of Understanding on the Conservation of Migratory Birds of Prey in Africa and Eurasia (Raptor MoU),[141] the Bern Convention, the Ramsar Convention and other relevant multilateral environmental agreements to fulfil their obligations with regard to avoiding or mitigating possible negative impacts of energy sector developments on migratory species.[142]

MoP7.Record.Annex8-Res7.5-WindturbinesandBatPopulations_adopted.pdf> accessed 7 September 2017.

137 Agreement on the Conservation of Small Cetaceans in the Baltic, North East Atlantic, Irish and North Seas, 1992 1772 *UNTS* 217. See also ASCOBANS Res 6.2 (18 September 2009) available at <http://www.ascobans.org/sites/default/files/document/MOP6_2009-2_UnderwaterNoise_1.pdf> accessed 7 September 2017.

138 Agreement on the Conservation of Cetaceans in the Black Sea, Mediterranean Sea and Contiguous Atlantic Area, 1996, 2183 *UNTS* 303 and 2003 amendments, UKTS No. 38 (2016). See also ACCOBAMS Res 4.17 (2010) ACCOBAMS-MOP4/2010/Res.4.17, available at <http://www.accobams.org/new_accobams/wp-content/uploads/2016/06/ACCOBAMS_MOP4_Res.4.17.pdf> accessed 7 September 2017.

139 Convention on the Conservation of European Wildlife and Natural Habitats, 1979, 1284 UNTS 210. See also Recommendation No. 109 (2004), Bern Convention Rec. 109, available at <https://wcd.coe.int/ViewDoc.jsp?p=&id=1493457&Site=&BackColorInternet=B9BDEE&BackColorIntranet=FFCD4F&BackColorLogged=FFC679&direct=true> accessed 7 September 2017.

140 The Task Force met for the first time from 1–2 December 2016 in South Africa. For the text of the Cape Town Declaration on reconciling renewable energy and power line development with migratory species conservation, see <http://www.cms.int/sites/default/files/document/ETF1_CapeTownDeclaration.pdf> accessed 7 September 2017.

141 The text of the Memorandum of Understanding on the Conservation of Migratory Birds of Prey in Africa and Eurasia is available at <http://www.cms.int/raptors/en/page/agreement-text> accessed 7 September 2017.

142 CMS Res 11.27, n. 128, annex.

5 Social Dimension

Offshore renewable energy sources may be a viable and sustainable solution for coastal communities that have limited or no access to modern energy services.[143] However, one of the greatest technical problems that must be overcome lies in the fact that the energy generated by offshore wind farms or wave, tide, salinity or thermal devices must be cabled to shore and connected to the existing energy grids. Moreover, renewable energy equipment used in that regard must also conform to certain standards of grid voltage, frequency and waveform purity requirements so that its benefits can be accessed by remote communities.[144]

Where offshore renewable energy projects are sited close to densely populated areas, proximity to major electrical load centres can significantly reduce the cost of transmission and offset transmission congestion.[145]

Planning, construction, and maintenance of offshore renewable energy operations also have the potential to create direct and indirect employment in various sectors including manufacturing, construction, operation and maintenance.[146] As the industry continues to grow, it has the potential to add thousands of new jobs. Consistent and effective policies in the field of education and training are crucial to support the renewables sector by providing the skills necessary for renewable energy deployment.[147] Important policy measures include strategic planning for skill needs; financing for renewable energy education, training and research; and the inclusion of renewable energy in educational programmes.[148] Equally important are the availability of comprehensive rules, regulations and standards that ensure the health, safety and decent working conditions for those employed in the offshore renewable energy industry. Most of those rules, regulations and standards are developed at the national level, with the exception of those applying to persons working or being transported on board ships.[149]

143 UNGA 'Report of the Secretary-General. Oceans and the law of the sea', n. 19, p. 21.

144 Ibid.; see also World Bank, 'Transmission expansion for renewable energy scale-up: emerging lessons and recommendations' (June 2011) Energy and Mining Sector Board Discussion Paper No. 26.

145 *The First Global Integrated Marine Assessment*, n. 19, p. 359.

146 Ibid.

147 IRENA, *Renewable Energy and Jobs* (2013), p. 80.

148 Ibid.

149 For further details, see section 9 on 'Ships, including the People on Board' below.

Notwithstanding the socioeconomic benefits of offshore renewable energy, the deployment of wind farms and other installations for the production of offshore renewable energy has also been perceived as a challenge by local communities.[150] The effect of impaired visual amenity from offshore renewable energy deployment can affect property prices, result in loss of recreational value, and reduce demand for tourism in coastal areas.[151] It can also affect historic and culturally significant resources.[152] These impacts are expected to be more prominent for an offshore wind project than for a wave, tidal or ocean current installation, as the latter will be underwater and of smaller scale for similar capacity.[153]

It is thus very important to involve local communities directly in identifying the sites for the installation of generators of offshore renewable energy, as well as the landing sites of related cables, and assessing benefits and costs connected with them.[154] Dissemination of information and education of stakeholders are crucial for a meaningful involvement of local communities in this decision-making process.[155] Environmental impact assessments, marine spatial planning and other management tools can facilitate broad and transparent stakeholder engagement in the planning of offshore renewable energy projects.

6 Offshore Renewable Energy Artificial Islands, Installations and Structures and Devices

Various technologies are used for the production of offshore renewable energy. These include artificial islands, installations and structures and devices installed either on the surface or underwater. Legal certainty is also particularly important for the development of and investment in such technology. This section will focus predominantly on offshore wind technology since it is currently most widely used.

150 UNGA 'Report of the Secretary-General. Oceans and the law of the sea', n. 19, p. 24.
151 *The First Global Integrated Marine Assessment*, n. 19, p. 359.
152 Ibid.
153 Ibid.
154 UNGA 'Report of the Secretary-General. Oceans and the law of the sea', n. 19, p. 24.
155 Ibid.

Offshore wind turbine installations and structures can either be fixed (through bottom-mounting) or floating (through the use of spar buoys, tension-leg platforms or semi-submersibles).[156] Hybrid systems use a combination of these three designs.[157] Currently, most structures are fixed. Until 2007, offshore wind turbines were installed in water depths below 20 metres and closer than 30 kilometres from shore. By contrast, turbines today are being installed routinely in water depths up to 40 metres and as far as 80 kilometers from shore.[158] Floating installations and structures are expected to start to be commercially available by 2020 and may offer a lower-cost alternative to fixed-bottom foundations due to the potential for standardization of foundation designs, maximization of onshore activity and the use of low-cost, readily available installation vessels.[159]

Most early offshore wind projects were built without an offshore substation. However, offshore wind farms located far from the shore require offshore substations in order to bring together all the energy produced by the wind turbines for conversion by transformers to a high voltage transmission. Offshore substation platforms are mounted on steel jacket structures or monopoles and house the transformers and electrical equipment required to collect, transform and export the power generated by turbines into transmission grids.[160] High voltage direct current (HVDC) power cables are used for grid connections greater than from about 80 to about 150 kilometres offshore.[161] Offshore HVDC converter stations are more complex, contain a lot more equipment and can have a mass greater than 10,000 tonnes.[162] The heaviest HVDC platform currently deployed is designed to connect three wind farms in the North Sea to the onshore grid, is rated at 900 MW and has a mass of about 20,000 tonnes.[163] Indeed, increased attention has been focused on connecting offshore HVDC substations to international or interstate HVDC supergrids.[164]

156 IRENA, *Innovation Outlook, Offshore Wind, Summary for Policy Makers*, 2016, p. 6, available at <http://www.irena.org/DocumentDownloads/Publications/IRENA_Innovation_Outlook_Offshore_Wind_2016_summary.pdf> accessed 7 September 2017.

157 A.N. Robertson and J.M. Jonkman, 'Load Analysis of Several Offshore Floating Wind Turbine Concepts', in ISOPE 1, Proceedings of the 21st International Offshore and Polar Engineering Conference, 2011, pp. 443–450.

158 IRENA, n. 156, p. 2.

159 Ibid., p. 5.

160 A. Chircop and P. L'Esperance, n. 62.

161 IRENA, *Innovation Outlook: Offshore Wind*, 2016, p. 37.

162 IRENA, n. 160, p. 37.

163 Ibid.

164 Ibid., p. 5.

The North Sea Countries Offshore Grid Initiative[165] cuts across the maritime zones of a number of States and aims to establish a shared electricity grid across the North Sea for harvesting its wind energy by aggregating the electricity output of the several production sites, thus enabling not only renewable energy technologies to be applied further offshore, but also more effective and efficient trading of energy between producers.[166] The Political Declaration on energy cooperation between the North Seas countries, signed on 6 June 2016,[167] includes the North Sea Wind Power Hub project. It is expected to be carried out by a multi-party consortium of energy companies and will aim to connect multiple wind-farms and solar panels spanning Dogger Bank, a large sandbank about 100 kilometres off the east coast of the United Kingdom to a central artificial island power hub in the North Sea from where generated wind energy can be distributed and transmitted over HVDC power lines to Belgium, Denmark, Germany, Norway, the Netherlands and the United Kingdom.[168]

UNCLOS sets out the rights and duties of States in relation to artificial islands, installations and structures in the different maritime zones.[169] In addition, the Convention's provisions have also been further developed by other legal instruments adopted at global and regional levels.

165 Memorandum of Understanding on the North Seas Countries Offshore Grid Initiative, signed on 3 December 2010, available at <http://www.benelux.int/files/8113/9625/9202/MoU_NSCOGI.pdf> accessed 7 September 2017. For further information on the Initiative, see <http://www.benelux.int/nl/kernthemas/holder/energie/nscogi-2012-report> accessed 7 September 2017.

166 V. Roeben, 'Governing Shared Offshore Electricity Infrastructure in the Northern Seas', *International and Comparative Law Quarterly*, vol. 62(4), 2013, pp. 840–841.

167 *Political Declaration on energy cooperation between the North Seas Countries*, signed on 6 June 2016, available at <http://www.benelux.int/files/9014/6519/7677/Political_Declaration_on_Energy_Cooperation_between_the_North_Seas_Countries.pdf> accessed 7 September 2017.

168 *Daily Mail*, 'Welcome to Dogger Island: How 'crazy' artificial power island in the North Sea could supply renewable energy to 80 million people in Europe by 2050' (8 March 2017), available at <http://www.dailymail.co.uk/sciencetech/article-4293250/Dutch-Danish-grid-operators-form-offshore-power-hub-North-Sea.html> accessed 7 September 2017; M. Lempriere, 'Could a daisy chain of artificial islands be the future of offshore wind?' (10 July 2010), available at <http://www.power-technology.com/features/featurecould-a-daisy-chain-of-artificial-islands-be-the-future-of-offshore-wind-5864912/> accessed 7 September 2017.

169 A detailed analysis of the legal regime in UNCLOS applicable to offshore wind farms, is provided by, for example, C. Fink, *The International Regulation of Offshore Wind Farms: under the 1982 Law of the Sea Convention (UNCLOS)*, University of Oslo, 2005, available at <https://www.duo.uio.no/handle/10852/20424> accessed 7 September 2017.

In internal waters, archipelagic waters and the territorial sea, coastal and archipelagic States exercise sovereignty, subject to the rights of navigation and overflight of other States. Article 60 of UNCLOS grants the coastal State the exclusive right to construct and to authorize and regulate the construction, operation and use of artificial islands; installations and structures for the purposes provided for in Article 56, which include the production of energy from water, currents and wind, and other economic purposes; and installations and structures which may interfere with the exercise of the rights of the coastal State in the zone.[170] The coastal State has exclusive jurisdiction over such artificial islands, installations and structures, including jurisdiction with regard to customs, fiscal, health, safety and immigration laws and regulations. Article 80 of UNCLOS provides that Article 60 applies *mutatis mutandis* to artificial islands, installations and structures on the continental shelf.[171]

UNCLOS clarifies that artificial islands, installations and structures do not possess the status of islands. They have no territorial sea of their own, and their presence does not affect the delimitation of the territorial sea, the EEZ or the continental shelf.[172]

UNCLOS does not define the terms 'artificial islands, installations and structures'. During the Third United Nations Conference on the Law of the Sea, the suggestion by its Drafting Committee to define 'installations' as including 'artificial islands and structures' was not accepted.[173] Recent discussions in the IMO suggest that States consider artificial islands, installations and structures to include not only fixed offshore oil platforms, floating production units, floating storage units, and mobile drilling units when secured to the bottom or held in position by other means and engaged in drilling, exploration or exploitation activities, but also offshore renewable energy installations.[174]

6.1 *Placement*

UNCLOS generally leaves the choice of placement of artificial islands, installations and structures to the coastal and archipelagic States subject to their

---

170 UNCLOS, n. 10, Article 60, para. 1.

171 Ibid., Article 80.

172 Ibid., Article 60, para. 8.

173 A/CONF.62/L.57/Rev.1 (1980), section VII, 'Items under consideration', para. 1, XIV *Official Records of the Third United Nations Conference on the Law of the Sea* at pp. 114 and 119 (Chairman, Drafting Committee). See also S.N. Nandan and S. Rosenne (eds.), *United Nations Convention on the Law of the Sea: A Commentary*, vol. III, Martinus Nijhoff, 1995), p. 151.

174 See subsection on 'Safety Zones' below.

obligations relating to the protection and preservation of the marine environment, and the obligation to have due regard to the rights and duties of other States, which include those relating to navigation and overflight and the laying of submarine cables and pipelines. UNCLOS specifically provides that artificial islands, installations and structures may not be established where interference may be caused to the use of recognized sea lanes essential to international navigation.[175]

The rights and duties with respect to navigation and overflight in the various maritime zones are set out in UNCLOS. This section will focus mainly on the regime in UNCLOS relating to navigation. In areas of internal waters enclosed by straight baselines which had not previously been considered as such,[176] in archipelagic waters[177] and in the territorial sea,[178] ships of all States enjoy the right of innocent passage,[179] unless rights of transit passage and archipelagic sea lanes passage apply. Passage by a foreign ship is considered non-innocent if it engages in any act aimed at interfering with any installations of the coastal State.[180] The coastal State and archipelagic State may adopt laws and regulations, in conformity with the provisions of UNCLOS and other rules of international law, relating to innocent passage, in respect of *inter alia* the safety of navigation and regulation of maritime traffic and the protection of installations and of cables and pipelines.[181] A coastal State and archipelagic State may, where necessary having regard to the safety of navigation, require foreign ships exercising the right of innocent passage to use such sea lanes and traffic separation schemes as it may designate or prescribe for the regulation of the passage of ships taking into account *inter alia* the recommendations of the competent international organization, that is the IMO.[182] In practice, States tend to submit any proposed ships' routeing measure to the IMO for adoption. The coastal State and archipelagic State must clearly indicate designated sea lanes and prescribed traffic separation schemes on charts to which due publicity must be given.[183] It must also give appropriate publicity to any danger to navigation of which it has knowledge.[184]

175 UNCLOS, n. 10, Article 60, para. 7.
176 Ibid., Article 8.
177 Ibid., Part IV.
178 Ibid., Part II, sections 1 to 3.
179 Ibid., Articles 8, 52 and Part II, section 3.
180 Ibid., Article 19, para. 2(k) and Article 52.
181 Ibid., Article 21, para. 1(b) and (c) and Article 52.
182 Ibid., Article 22, paras. 1 to 3 and Article 52.
183 Ibid., Article 22, para. 4 and Article 52.
184 Ibid., Articles 24 and 52.

In straits used for international navigation between one part of the high seas or an EEZ and another part of the high seas or an EEZ and in archipelagic sea lanes, all ships and aircraft have the rights and duties with respect to transit passage and archipelagic sea lanes passage as provided for in UNCLOS.[185] States bordering such straits and archipelagic States may designate sea lanes and prescribe traffic separation schemes for navigation in straits used for international navigation and in archipelagic sea lanes only if they have been adopted by the competent international organization,[186] that is the IMO.

Foreign ships exercising the right of innocent passage must comply with the laws and regulations adopted by the coastal State and archipelagic State and with generally accepted international regulations relating to the prevention of collisions at sea.[187] Such regulations are contained in the Convention on the International Regulations for Preventing Collisions at Sea[188] (COLREGS). When exercising transit and archipelagic sea lanes passage, foreign ships must comply with generally accepted international regulations, procedures and practices for safety at sea, including COLREGS, the 1974 International Convention for the Safety of Life at Sea[189] (SOLAS 1974) and its Protocol of 1988 and the General Provisions on Ships' Routeing,[190] as amended, as well as the laws and regulations adopted by the States bordering straits and archipelagic States.[191]

UNCLOS does not contain specific provisions concerning the designation of sea lanes and traffic separation schemes for the purpose of safety of navigation in the EEZ. However, it requires States to promote the adoption, wherever appropriate, of routeing systems designed to minimize the threat of accidents which might cause pollution of the marine environment, including the coastline, and pollution damage to the related interests of coastal States, acting through competent international organizations or general diplomatic

185 Ibid., Articles 38, 39, 40, 41, 42, 43, 44 and 53 and 54.

186 Ibid., Articles 41 and 53, respectively.

187 Ibid., Article 21, para. 4 and Article 52; IMO, 'Implications of the United Nations Convention on the Law of the Sea for the International Maritime Organization. Study by the Secretariat of the International Maritime Organization (IMO)' (30 January 2014) IMO Doc LEG/MISC.8, pp. 31–32, available at <http://www.imo.org/en/OurWork/Legal/Documents/LEG%20MISC%208.pdf> accessed 7 September 2017.

188 Convention on the International Regulations for Preventing Collisions at Sea, 1972, 1050 UNTS 17.

189 International Convention for the Safety of Life at Sea, 1184 UNTS 277, [SOLAS].

190 IMO Assembly Res A.572(14) (20 November 1985), as amended.

191 UNCLOS, n. 10, Articles 39, 40, 41, 42 and 53 and 54; IMO, n. 187, pp. 32–34.

conference.[192] IMO has clarified that the General Provisions on Ships' Routeing also apply beyond the territorial sea.[193]

Since the General Provisions on Ships Routeing provide guidelines only for the establishment of single objects, such as e.g. drilling rigs (MODUs), exploration platforms and other similar structures, but not for multiple structures, such as wind farms, which may have a different impact on the ability of vessels to act according to COLREGs, compared to single objects, the IMO Maritime Safety Committee adopted an amendment to the General Provisions on Ships' Routeing in November 2016,[194] subject to subsequent confirmation by the IMO Assembly at its session in 2017.[195] The amendment has the effect of adding a new paragraph on establishing multiple structures at sea. It recommends that '[i]n planning to establish multiple structures at sea, including but not limited to wind turbines, governments should take into account, as far as practicable, the impact these could have on the safety of navigation, including any radar interference. Traffic density and prognoses, the presence or establishment of routeing measures in the area, and the manoeuvrability of ships and their obligations under the 1972 Collision Regulations should be considered when planning to establish multiple structures at sea. Sufficient manoeuvring space extending beyond the side borders of traffic separation schemes should be provided to allow evasive manoeuvres and contingency planning by ships making use of routeing measures in the vicinity of multiple structure areas'.[196]

On the high seas, all States enjoy the freedom to construct artificial islands and other installations permitted under international law, subject to Part VI of UNCLOS and the duty of due regard for the interests of other States and the rights under the Convention with respect to activities in the Area.[197]

UNCLOS requires activities in the Area and other activities in the marine environment to be carried out with reasonable regard for each other. Article 147 provides for the possibility of installations being used for carrying out activities in the Area and used exclusively for peaceful purposes. As mentioned above, there are different views regarding the meaning of 'resources' and thus the applicability of Part XI of UNCLOS to installations used for offshore

192 UNCLOS, n. 10, Article 211, para. 1.

193 IMO, n. 187, p. 34.

194 IMO, Res MSC.419(97) (6 December 2016) MSC 97/22/Add.1, annex 12.

195 The 30th session of the IMO Assembly will take place from 27 November to 6 December 2017.

196 IMO, 'Report to the Maritime Safety Committee' (22 March 2016) IMO Doc NCSR 3/29, para. 4.7 and annex 5.

197 UNCLOS, n. 10, Article 87.

renewable energy.[198] Article 147 requires installations in the Area to be erected, emplaced and removed solely in accordance with Part XI and subject to the rules, regulations and procedures of the International Seabed Authority. Installations may not be established where interference may be caused to the use of recognized sea lanes essential to international navigation or in areas of intense fishing activity. Safety zones must be established around installations with appropriate markings to ensure the safety of both navigation and the installations. The configuration and location of such safety zones must not be such as to form a belt impeding the lawful access of shipping to particular maritime zones or navigation along international sea lanes.

UNCLOS requires due notice to be given of the construction of an artificial island, installation and structure in the EEZ and on the continental shelf and of its erection, emplacement or removal in the Area, and permanent means for giving warning of their presence must be maintained.[199]

Since offshore wind installations at sea may pose a risk to air traffic, such warning would also need to comply with the standards developed by the International Civil Aviation Organization (ICAO) setting out requirements for the marking and/or lighting of wind turbines over a certain height.[200] ICAO has also developed a number of recommendations on how to mark and light such turbines.[201] Of relevance are also the recommendations on the marking of man-made offshore structures of the International Association of Marine Aids to Navigation and Lighthouse Authorities, which include guidance on the marking of offshore wind farms and offshore wave and tidal energy devices.[202]

6.2 *Safety Zones*

In its Articles 60 and 80, UNCLOS grants the coastal State the discretion to, where necessary, establish reasonable safety zones around artificial islands,

198 See section 3 on 'States' Rights to Produce Offshore Renewable Energy in the Various Maritime Zones' above.

199 UNCLOS, n. 10, Articles 60, 80 and 147.

200 See Convention on International Civil Aviation, Annex 14, chapter 6.2.4, *ICAO International Standards and Recommended Practices, Annex 14 to the Convention on International Civil Aviation: Aerodromes – Volume I: Aerodrome Design and Operations*, Seventh Edition, 2016, p. 6-11.

201 Ibid., pp. 6-11 to 6-12.

202 IALA Recommendation O-139 'The Marking of Man-Made Offshore Structures, Edition 2' (December 2013), available at <https://www.transportstyrelsen.se/globalassets/global/sjofart/dokument/sjotrafik_dok/marking-of-man-made-offshore-structures.pdf> accessed 7 September 2017.

installations and structures in which it may take appropriate measures to ensure the safety both of navigation and of the installation and structure. Safety zones may not be established where this may cause interference with the use of recognized sea lanes essential to international navigation. The breadth of the safety zones shall be determined by the coastal State, taking into account applicable international standards. Safety zones must not exceed a distance of 500 metres around them, measured from each point of their outer edge, except as authorized by generally accepted international standards or as recommended by the competent international organization. Due notice must be given of the extent of safety zones. All ships must respect these safety zones and comply with generally accepted international standards regarding navigation in the vicinity of artificial islands, installations and structures and safety zones.

Article 147 provides for the establishment of safety zones around installations used for carrying out activities in the Area and their appropriate markings to ensure the safety of both navigation and the installation. UNCLOS prescribes that the configuration and location of such safety zones must not be such as to form a belt impeding lawful access to shipping to particular maritime zones or navigation and international sea-lanes.

Guidelines on Safety Zones and Safety of Navigation around Offshore Installations and Structures were adopted by IMO in 1989[203] and 2010.[204] The 1989 Guidelines are focused on installations and structures engaged in the exploitation of natural resources on the continental shelf and in the EEZ, while the 2010 Guidelines also cover artificial islands, installations and structures used for the production of offshore renewable energy. The Guidelines acknowledge that some offshore artificial islands, installations and structures are complex systems that present particular challenges for safe navigation and request governments *inter alia* to implement the recommendations in the 1989 Guidelines and to take appropriate measures to ensure navigation charts clearly reflect the location and projected swing or movement, if any, with the wind and seas of Floating Production Storage Offloading units, including their connected associated and necessary structures, installations, vessels, shuttle tankers and/or tugs in its operations, and other similarly situated installations or structures, that rotate around a fixed mooring.[205]

203 IMO Assembly Res A.671(16) (19 October 1989).

204 IMO, 'Guidelines for safety zones and safety of navigation around offshore installations and structures' (7 December 2010) IMO Doc SN.1/Circ.295.

205 Ibid.

The 2010 Guidelines were developed by the then IMO Sub-Committee on Safety of Navigation (NAV)[206] following its consideration of a proposal to amend the General Provisions on Ships Routeing to add guidelines for the adoption and designation of safety zones larger than 500 metres around artificial islands, installations and structures in the EEZ including multiple structure installations, as well as the report of the Correspondence Group that had been mandated to work on the issue intersessionally.[207] The Correspondence Group had developed a draft SN circular and also draft guidelines for consideration of requests for safety zones greater than 500 metres around artificial islands, installations and structures in the EEZ.[208] However, the NAV Sub-Committee decided that there was no demonstrated need at present to establish safety zones greater than 500 metres around artificial islands, installations and structures in the EEZ or to develop guidelines to do so.[209] Noteworthy is the definition of artificial islands, installations and structures in the draft guidelines that were developed by the Correspondence Group to include 'fixed platforms, floating production units, floating storage units, mobile drilling units when secured to the bottom or held in position by other means and engaged in drilling, exploration or exploitation activities, and offshore renewable energy installations including wind, wave and tidal turbine generators'.[210] Multiple structure installations are defined in the draft guidelines as including 'multiple artificial islands, installations and structures, as defined above, and offshore renewable energy installations including wind farms and wave or tidal arrays planned in such density that safe navigation between them cannot be maintained'.[211]

6.3 *Removal and Disposal*

UNCLOS Article 60 requires any installations or structures which are abandoned or disused to be removed to ensure safety of navigation, taking into account any generally accepted international standards established in this

---

206 It was renamed by IMO as the Sub-Committee on Navigation, Communications and Search and Rescue.

207 IMO, 'Guidelines for consideration of requests for safety zones larger than 500 metres around artificial islands, installations and structures in the EEZ: Report of the Correspondence Group, Submitted by the United Kingdom' (23 April 2010) IMO Doc. NAV 56/4.

208 Ibid., annexes 1 and 2.

209 IMO, 'Report to the Maritime Safety Committee' (31 August 2010) IMO Doc NAV 56/20, para. 4.15.

210 IMO, no. 207, annex 1, para. 3.2.

211 Ibid., para. 3.3.

regard by the competent international organization. Such removal shall also have due regard to fishing, the protection of the marine environment and the rights and duties of other States. Appropriate publicity must be given to the depth, position and dimensions of any installations or structures not entirely removed.[212]

In 1989, IMO Assembly adopted Guidelines and Standards for the Removal of Offshore Installations and Structures on the Continental Shelf and in the Exclusive Economic Zone bearing in mind Article 60 of UNCLOS.[213]

The Guidelines define the standards to be followed by the coastal State when making decisions regarding the removal of abandoned or disused installations and structures. Abandoned offshore installations are required to be removed, except in certain cases. A decision to allow an installation to remain, in whole or in part, on the seabed should take into account the circumstances described in the resolution. The Guidelines and Standards also incorporate and extend the requirement under Article 60 (3) of UNCLOS to provide 'appropriate publicity' for partial removal. Notification not only of partial removal but also of non-removal should be forwarded to the IMO. That Organization may establish that the publicity requirement takes into account the depth, position and dimension of the installations and structures not entirely removed, as provided in Article 60 of UNCLOS. The Guidelines provide that from 1 January 1998 onward, no installation or structure should be placed on any continental shelf or in any EEZ unless the design and construction of the installation or structure is such that entire removal upon abandonment or permanent disuse would be feasible.

Where removal of an installation is to be solved by dumping at sea, the provisions of UNCLOS and those of the 1972 Convention on the Prevention of Marine Pollution by Dumping of Wastes and Other Matter (London Convention)[214] and the 1996 Protocol to the Convention on the Prevention of Marine Pollution by the Dumping of Wastes and Other Matter, 1972[215] (1996 Protocol) apply.

---

212 UNCLOS, n. 10, Article 60, para. 3.

213 IMO Assembly Res A.672(16) (19 October 1989), available at <http://www.imo.org/blast/blastDataHelper.asp?data_id=22503&filename=A672(16)E.pdf> accessed 7 September 2017.

214 Convention on the Prevention of Marine Pollution by Dumping of Wastes and Other Matter, 1972, 1046 UNTS 138 ['London Convention'].

215 1996 Protocol to the Convention on the Prevention of Marine Pollution by the Dumping of Wastes and Other Matter, 36 ILM 1 (1997) ['1996 Protocol'].

UNCLOS requires States to endeavour to establish global and regional rules and standards and recommended practices and procedures to prevent, reduce and control pollution by dumping, acting through competent international organizations or diplomatic conference. National laws, regulations and measures must be no less effective in preventing, reducing and controlling such pollution than the global rules and standards.[216] Such laws and regulations and applicable international rules and standards are required to be enforced by States.[217] The global and regional framework which has been established in this regard consists of several treaties and other legal instruments.[218] Global rules and standards are provided in the London Convention, as periodically amended by decisions of its Contracting Parties, and the 1996 Protocol which comprehensively and substantially amended the parent convention.

The deliberate disposal of platforms or other man-made structures at sea is included within the definition of dumping in UNCLOS[219] which had incorporated the definition of 'dumping' contained in the London Convention. The latter definition was further elaborated on in the 1996 Protocol to include: 'any abandonment or toppling at site of platforms or other man-made structures at sea, for the sole purpose of deliberate disposal'.[220]

Both the London Convention[221] and its 1996 Protocol,[222] provide for the possibility of offshore installations to be disposed of at sea with a permit in accordance with the Specific Guidelines for Assessment of Platforms or other Man-made Structures at Sea adopted in 2000.[223] However, the 1996 Protocol

216 UNCLOS, n. 10, Article 210.

217 Ibid., Article 216.

218 These include at the regional level, OSPAR Decision 98/3 (1998) 'Disposal of Disused Offshore Installations'; 'Guidelines – Dumping of Platforms and other man-made structures at sea' adopted by the Contracting Parties to the Convention for the Protection of the Mediterranean Sea against Pollution in 2003 (1 October 2003).

219 UNCLOS, n. 10, Article 1, para. 5(a)(ii).

220 IMO, n. 187, p. 56.

221 London Convention, n. 214, Article III.1(a)(ii).

222 1996 Protocol, n. 215, Article 1.4.1.2.

223 IMO, 'Report of the twenty-second Consultative Meeting of Contracting Parties to the London Convention' (25 October 2000) IMO Doc LC 22/14, annex 7. Revisions to the Guidelines are being considered by the Scientific Group of the London Convention and of the 1996 Protocol, see IMO, 'Report of the fortieth Meeting of the Scientific Group of the London Convention and the eleventh Meeting of the Scientific Group of the London Protocol' (24 April 2017) IMO Doc LDC/SG 40/16, paras. 2.38–2.50 and annex 4.

requires Contracting Parties to pay particular attention to opportunities to avoid dumping in favour of environmentally preferable alternatives.[224]

6.4 *Protection and Preservation of the Marine Environment*

There are potential environmental benefits from the installation of offshore artificial islands, installations, structures and devices since they can provide marine organisms with artificial reefs, also investigations have found greater fish abundance in the vicinity of offshore turbines compared to surrounding areas.[225] However, special care is needed in siting installations that can affect migration routes or feeding, breeding or nursery areas.[226] Also noise created during pile-driving operations involves sound pressure levels that are high enough to impair hearing in marine mammals and disrupt their behaviour even at a considerable distance from the construction site.[227] Fixed and moving parts of the artificial islands, installations, structures and devices can lead to fatal strikes or collisions with birds and aquatic fauna.[228]

The measures that States are required to take to pursuant to UNCLOS Part XII on the protection and preservation of the marine environment must deal with all sources of pollution of the marine environment. These include measures designed to minimize to the fullest possible extent pollution from installations and devices used in the exploration or exploitation of the natural resources of the seabed and subsoil,[229] and pollution from other installations and devices operating in the marine environment, in particular measures for preventing accidents and dealing with emergencies, ensuring the safety of operations at sea, and regulating the design, construction, equipment, operation and manning of such installations or devices.[230] As previously indicated, States are also obligated under UNCLOS and customary international law to conduct environmental impact assessments.

In addition, UNCLOS requires coastal States to adopt laws and regulations to prevent, reduce and control pollution of the marine environment arising from or in connection with seabed activities subject to their jurisdiction and from artificial islands, installations and structures under their jurisdiction which must be no less effective than international rules, standards and

224 1996 Protocol, n. 215, Article 4.1.2.
225 *The First Global Integrated Marine Assessment*, n. 19, p. 357.
226 Ibid., p. 27.
227 Ibid., p. 358.
228 Ibid., see also section 4 above on 'The Environmental Dimension'.
229 UNCLOS, n. 10, Article 194, para 3.
230 Ibid.

recommended practices and procedures to be established especially through competent international organizations or diplomatic conference.[231]

States must also adopt laws and regulations to prevent, reduce and control pollution of the marine environment from activities in the Area undertaken by vessels, installations, structures and other devices flying their flag or of their registry or operating under their authority, as the case may be. Such laws and regulations must be no less than effective than the international rules, regulations and procedures that are to be established in accordance with Part XI of UNCLOS.[232]

No global treaty has been developed which specifically addresses the prevention, reduction and control of pollution of the marine environment from artificial islands, installations and structures. However, global rules and standards have been developed by IMO to address pollution from ships which also apply to fixed or floating platforms. In particular, the 1973 International Convention for the Prevention of Pollution from Ships as modified by the Protocol of 1978[233] (MARPOL), which includes regulations aimed at preventing and minimizing pollution from ships, both accidental and from routeing operations,[234] applies to pollution from 'fixed or floating platforms' other than pollution resulting from the 'release of harmful substances directly arising from the exploration, exploitation and associated offshore processing of seabed mineral resources'.[235]

Measures relating to marine biofouling are contained in the 2001 International Convention on the Control of Harmful Anti-Fouling Systems on Ships.[236] The Convention prohibits the use of harmful organotins in anti-fouling paints used on ships, including fixed or floating platforms,[237] and establishes a mechanism to prevent the potential future use of other harmful substances in anti-fouling systems.

---

231 Ibid., Article 208.

232 Ibid., Article 209.

233 International Convention for the Prevention of Pollution from Ships, 1973, as modified by the Protocol of 1978, 1340 *UNTS* 62 ['MARPOL'].

234 For a description of MARPOL, see IMO website <http://www.imo.org/en/about/conventions/listofconventions/pages/international-convention-for-the-prevention-of-pollution-from-ships-(marpol).aspx> accessed 7 September 2017.

235 MARPOL, n. 233, Article 2.

236 IMO, 'Adoption of the Final Act of the Conference and any instruments, recommendations and resolutions resulting from the work of the Conference: International Convention on the Control of Harmful Anti-fouling Systems on Ships, 2001' (18 October 2001) AFS/CONF/26.

237 Ibid., Article 2(9).

The 1990 International Convention on Oil Pollution Preparedness Response and Cooperation,[238] while not including offshore units in its definition of ships, nonetheless requires operators of offshore units under the jurisdiction of Parties to have oil pollution contingency plans or similar arrangements in place.[239]

The definition of dumping in UNCLOS and in the London Convention and its 1996 Protocol includes any deliberate disposal of wastes or other matter from platforms or other man-made structures at sea.[240] The 1996 Protocol also defines dumping to include any storage of wastes or other matter in the seabed and the subsoil thereof from platforms or other man-made structures at sea.[241] UNCLOS requires national laws, regulations and measures to prevent, reduce and control pollution by dumping to be no less effective than the global rules and standards to be established.[242] Such rules and standards are provided in the London Convention and the 1996 Protocol.

States are also required to take measures 'necessary to protect and preserve rare or fragile ecosystems as well as the habitat of depleted, threatened or endangered species and other forms of marine life'.[243] In that regard, the aforementioned global and regional instruments that contain measures to address potential or actual adverse impacts on marine biodiversity as result of activities involving offshore renewable energy technologies are particularly noteworthy.[244]

Notwithstanding the provisions in UNCLOS and the measures that have been adopted thus far, offshore installations and structures erected for the production of offshore renewable energy lack comprehensive regulation,[245] including in relation to responsibility and liability.[246] It has been suggested that priority attention should be directed to the development of clear procedures

---

238 International Convention on Oil Pollution Preparedness, Response and Cooperation, 1990, 1891 *UNTS* 78.

239 IMO, n. 187, p. 78.

240 UNCLOS, n. 10, Article 1, para. 5; London Convention, n. 214, Article III.1(a); 1996 Protocol, n. 215, Article 1.4.1.1.

241 1996 Protocol, n. 215, Article 1.4.1.3.

242 UNCLOS, n. 10, Article 210.

243 Ibid., Article 194, para. 5.

244 For further details, see section 4 above on 'The Environmental Dimension'.

245 H.K. Müller and M. Roggenkamp, 'Regulating offshore energy sources in the North Sea – Reinventing the wheel or a need for more coordination?', *The International Journal of Marine and Coastal Law*, vol. 29(4), 2014, pp. 716–737; M. Gavouneli, n. 84.

246 M. Gavouneli, n. 84.

and requirements for environmental impact assessments and strategic environmental assessments.[247]

7 Submarine Cables

As larger scale offshore renewable energy infrastructure has grown, so has the electricity infrastructure to collect this energy in a stable way for usage onshore. Submarine cables play a critical role in that regard since they transfer the power generated to shore. When offshore wind farms are located further offshore, alternating current (AC) electrical power from the windfarm is converted into high voltage, direct current (HVDC) power using an offshore platform mounted converter station. It then transmits the direct current through a HVDC power cable to a land-based electrical network where it is converted back again to AC by another HVDC converter station.[248] More efficient technology is under development which envisages repowering wind farms with HVDC nodes.[249] Increased attention is directed to connecting offshore HVDC substations to international HVDC supergrids,[250] as mentioned above.

The rights and duties of States in the various maritime zones with respect to power cables are set out in UNCLOS. Those provisions are based on the provisions in the 1958 Geneva Conventions on the High Seas, the Continental Shelf and the Territorial Sea. These in turn are based on the 1884 International Convention for the Protection of Submarine Telegraph Cables. However, the International Law Commission, in preparing draft articles concerning the law of the sea for the General Assembly, which later formed the basis of the Geneva Conventions, extended article II of the 1884 Convention to include pipelines and high voltage power cables.[251] The specific reference to high-voltage power cables in Article 113 of UNCLOS thus suggests that the general reference to submarine cables in other articles of UNCLOS is intended to include such cables.

---

247 D. Leary and M. Esteban, 'Climate Change and Renewable Energy from the Ocean and Tides: Calming the Sea of Regulatory Uncertainty', *International Journal of Marine and Coastal Law*, vol. 24, 2009, p. 644.

248 IRENA, n. 161, p. 36.

249 Ibid., pp. 76, 78.

250 Ibid., p. 5.

251 See Commentary of the International Law Commission on draft article 62 concerning the law of the sea, *Yearbook of the International Law Commission, 1956*, vol. II, document A/3169, chap. II, sect. III, pp. 293–294.

UNCLOS provides that the coastal State exercises sovereignty over its territorial sea[252] and has the right to establish conditions for cables entering its territory or territorial sea.[253] In archipelagic waters, over which an archipelagic State exercises sovereignty,[254] the existing submarine cables laid by other States and passing through its waters without making a landfall must be respected by the archipelagic State.[255] It is also required to permit the maintenance and replacement of such cables upon receiving due notice of their location and the intention to repair or replace them.[256]

On the continental shelf, the coastal State has jurisdiction over cables constructed or used in connection with the exploration of the continental shelf or exploitation of its resources or the operations of artificial islands, installations and structures under its jurisdiction,[257] which could include cables connecting wind farms to an offshore substation, and possibly an offshore electricity grid.[258] UNCLOS provides that the coastal State may not impede the laying or maintenance of submarine cables by other States. When exercising their right to lay such cables all States are required to have due regard to cables already in position.[259]

In the EEZ, the coastal State has jurisdiction with regard to the protection and preservation of the marine environment and exclusive jurisdiction over artificial islands, installations and structures. All States can exercise the high seas freedoms of navigation and of the laying of submarine cables, amongst others, and other internationally lawful uses of the sea related to those freedoms, such as those associated with the operation of ships and submarine cables.[260] UNCLOS applies by reference, in so far as they are not incompatible with Part V on the EEZ, *inter alia* the articles of the high seas setting out the rights and duties of States relating to submarine cables on the high seas.[261] In exercising their rights and in performing their duties, States are required to have due regard to the rights and duties of the coastal State and comply with the latter's laws and regulations adopted in accordance with UNCLOS and

252 UNCLOS, n. 10, Article 2.

253 Ibid., Article 79, para. 4.

254 Ibid., Article 49.

255 Ibid., Article 51, para. 2.

256 Ibid.

257 Ibid., Article 79, para. 4.

258 V. Roeben, n. 166, p. 847.

259 Ibid., Article 79, paras. 1, 5.

260 UNCLOS, n. 10, Article 58, para. 1.

261 Ibid., para. 2.

other rules of international law in so far as they are not incompatible with Part V.[262]

UNCLOS provides that on the bed of the high seas beyond the continental shelf, all States are entitled to lay submarine cables,[263] subject to Part VI of the Convention.[264] In exercising their rights and performing their duties States must have due regard to the interests of other States in their exercise of the freedom of the high seas and also for the rights under UNCLOS with respect to activities in the Area.[265] UNCLOS requires States to adopt the necessary laws and regulations to make it a punishable offence if a ship flying its flag or a person subject to its jurisdiction breaks or injures a submarine cable beneath the high seas wilfully or through culpable negligence, or breaks or injures a high-voltage power cable, or engages in conduct calculated or likely to result in such breaking or injury. Situations of *force majeure* are excluded, if all necessary precautions were taken to avoid such break or injury.[266] States must also have laws and regulations in place to provide that persons subject to their jurisdiction who are the owners of a submarine cable beneath the high seas must bear the cost of repairs if while laying or repairing a cable, a break or injury is caused to another submarine cable.[267] Finally, national laws and regulations must also provide for indemnification by the owner of the submarine cable for an anchor, net or other fishing gear that was sacrificed by the owners of ships to avoid injuring the cable, provided that the owner of the ship took all reasonable precautionary measures beforehand.[268]

As previously indicated, the different interpretations of Article 80 may also result in different views of the extent of coastal State's jurisdiction over cables used in connection with the operations of artificial islands, installations and structures on the continental shelf beyond 200 nautical miles from the baselines from which the breadth of the territorial sea is measured and on the continental shelf within 200 nautical miles where a proclaimed EEZ does not extend to 200 nautical miles or where such zone has not been proclaimed. Moreover, the extent of the rights of coastal States with respect to submarine cables located on the continental shelf which are connecting offshore substations with the onshore grid is not clear. Article 79 of UNCLOS has been interpreted to

262 Ibid., para. 3.
263 Ibid., Article 112.
264 Ibid., Article 87, para. 1 (c) and Article 79, para. 5.
265 Ibid., para. 2.
266 Ibid., Article 113.
267 Ibid., Article 114.
268 Ibid., Article 115.

mean that such cables would not fall under the jurisdiction of the coastal State since they do not serve to 'operate' specific wind farms,[269] leaving the coastal State in the same position as other States that are entitled to lay submarine cables on the continental shelf.[270] On the other hand, the absence of jurisdiction of the coastal State in the EEZ and on the continental shelf with respect to submarine cables connecting substations and the onshore grid would be in sharp contrast to the sovereign rights of the coastal State in the EEZ with regard to the production of wind energy and its exclusive jurisdiction over artificial islands, installations and structures, such as offshore wind farms. This issue would thus also benefit from clarification. In that regard, the possibility of future above water HVDC nodes connecting clusters of wind farms could also be considered.[271] Another jurisdictional issue that may require further clarification relates to the allocation of responsibility in the case of a multi-State transmission grid project.

Clarity on jurisdictional issues is also important for determining who is responsible for taking and enforcing measures to protect and preserve the marine environment. While UNCLOS does not include a specific provision addressing the protection and preservation of the marine environment from activities associated with the laying of submarine cables, the general obligation of States to protect and preserve the marine environment in UNCLOS[272] and other relevant provisions apply, including the requirement to take the measures 'necessary to protect and preserve rare or fragile ecosystems as well as the habitat of depleted, threatened or endangered species and other forms of marine life',[273] and to prevent, reduce and control pollution of the marine environment resulting from the use of technologies,[274] as well as the requirements regarding monitoring and environmental assessment.[275]

Cables can have a benign impact on the marine environment as they can act as artificial reefs.[276] However, submerged cables, in carrying electricity from offshore renewable energy technology to substations, emit low frequency electric and magnetic fields, which can potentially impact marine organisms as

269 V. Roeben, n. 166, pp. 847–848.
270 UNCLOS, n. 10, Article 79, para. 1.
271 V. Roeben, n. 166, p. 849.
272 UNCLOS, n. 10, Article 192.
273 Ibid., Article 194, para. 5.
274 Ibid., Article 196.
275 Ibid., Articles 204 to 206.
276 International Cable Protection Committee (ICPC), *About Submarine Power Cables* (2011), available at <https://www.iscpc.org/publications> accessed 7 September 2017.

these sources become more common.[277] Shielding can restrict directly emitted electric fields but cannot shield the magnetic field component.[278] Also to be considered are the increased temperature in the cable surface as a result of heat lost during the transport of energy and subsequent warming of the surrounding environment. Such changes in temperature can potentially impact benthic organisms.[279]

At the global level, the Guidelines on how to avoid or mitigate impacts on migratory species of the deployment of renewable energy technologies, including those used for ocean energy,[280] referred to earlier, which were endorsed by the Conference of the Parties to the CMS in 2014, recommend *inter alia* carrying out an environmental impact assessment and strategic environmental assessment, as well as burying undersea cables within the ocean energy development array and at the landfall connection to depths within the sediment that will minimize or eliminate the impacts from electromagnetic fields to sea turtles and marine mammals.[281]

The OSPAR Commission has undertaken an assessment of the environmental impacts of cables[282] and issued Environmental Practice in Cable Laying and Operation, which recommend the application of the precautionary principle and the implementation of appropriate mitigation measures, such as burying cables to avoid the impairment of marine species by electromagnetic fields and routeing cables away from protected or sensitive marine areas.[283] Some industry representatives have challenged those recommendations and suggested that the Guidelines be reviewed and amended in partnership with the industry.[284]

---

277 *The First Global Integrated Marine Assessment*, n. 19, p. 358.

278 Ibid.

279 Ibid. AC cables are theoretically likely to emit more heat than direct current cables. For further information, see A. Simcock, 'Submarine Cables and Pipelines', in *The First Global Integrated Marine Assessment*, n. 19, p. 282.

280 CMS Res 11.27, n. 128; and 'Renewable Energy Technologies and Migratory Species: Guidelines for Sustainable Deployment', n. 129.

281 CMS 'Renewable Energy Technologies and Migratory Species: Guidelines for Sustainable Deployment', n. 129, pp. 51–52.

282 OSPAR Commission, *Assessment of the environmental impacts of cables* (2009), available at <http://qsr2010.ospar.org/media/assessments/p00437_Cables.pdf> accessed 7 September 2017.

283 OSPAR Commission, Agreement 2012–2, n. 133; S. Mc Donald and D.L. VanderZwaag, n. 51, p. 323.

284 European Subsea Cables Association, 'European Subsea Cables Association position statement on OSPAR publication 'Guidelines on Best Environmental Practice (BEP) in

Highlighting its concerns, the industry has pointed to the damage caused to cables as a result of mainly fishing and anchoring activities[285] and the lack of national legislation to implement UNCLOS obligations to protect international cable infrastructure beyond territorial waters.[286] In that regard, it can be noted that the United Nations General Assembly has encouraged the adoption by States of laws and regulations addressing the breaking or injury of submarine cables or pipelines beneath the high seas done wilfully or through culpable negligence by a ship flying its flag or by a person subject to its jurisdiction, in accordance with international law, as reflected in the Convention.[287]

The industry has also raised the need for constructive interaction with other seabed uses, for example, the laying of telecommunications cables, bottom fishing, exploitation of minerals and hydrocarbons,[288] and shipping to prevent damage caused by anchors. The International Cable Protection Committee issued a Recommendation on 'The Proximity of Offshore Renewable Wind Energy Installations and Submarine Cable Infrastructure in National Waters' in 2013.[289] With respect to areas beyond national jurisdiction, the industry has indicated that submarine cables have a very small ecological footprint on the conservation and sustainable use of marine biological diversity.[290]

Disputes concerning the alleged contravention of the provisions of UNCLOS by a coastal State in regard to the freedom and rights of the laying of submarine cables or other internationally lawful uses of the sea specified in Article 58, are subject to compulsory procedures entailing binding decisions.[291]

Cable Laying and Operation" (April 2015), available at < http://www.escaeu.org/news/?newsid=42> accessed 7 September 2017.

285 Ibid.

286 ICPC, n. 276.

287 UNGA Res. 71/257, n. 7, para. 170.

288 ICPC, n. 276.

289 ICPC Recommendation No. 13 (Issue 2A, November 2013).

290 ICPC, 'Submarine Cables and BBNJ. Preparatory Committee established by General Assembly resolution 69/292: Development of an internationally binding instrument under the United Nations Convention of the Law of the Sea on the conservation and sustainable use of marine biological diversity of areas beyond national jurisdiction' (2016), available at <http://www.un.org/depts/los/biodiversity/prepcom_files/ICC_Submarine_Cables_&_BBNJ_August_2016.pdf> accessed 7 September 2017.

291 UNCLOS, n. 10, Article 297, para. 1(a).

8 Pipelines

Pipes are used for example in ocean thermal energy conversion (OTEC) in order to bring large volumes of colder water up from the depths while warmer water is also drawn into the OTEC plant from the ocean surface.

UNCLOS does not restrict the application of its provisions to a particular type of pipeline, although they are commonly perceived as pipelines transferring hydrocarbons. The legal regime in UNCLOS with respect to pipelines could thus also apply to pipelines used in OTEC. While the provisions of UNCLOS relating to submarine cables entering the territory or territorial sea,[292] in the EEZ and on the bed of the high seas beyond the continental shelf,[293] as described above are also applicable to pipelines,[294] as well as the provision relating to the settlement of disputes,[295] the provisions of UNCLOS relating to submarine cables on the continental shelf differ from those applicable to pipelines.[296] UNCLOS grants the coastal State broader jurisdiction with respect to the laying and maintenance of pipelines on its continental shelf. While the coastal State may not impede the laying or maintenance of pipelines on the continental shelf, such action is subject to its right to take reasonable measures for the exploration of the continental shelf, the exploitation of its natural resources and the prevention, reduction and control of pollution from pipelines.[297] In addition, the delineation of the course of the laying of a pipeline on the continental shelf is subject to its consent.[298] In archipelagic waters, there is no requirement for the archipelagic State to respect existing pipelines as in the case of submarine cables.

While it is certain that physical, chemical and biological impacts will occur during the installation and operation of an OTEC facility, the magnitude and extent of these impacts are not known. Impacts to biota include nutrient redistribution, entrainment and impingement, organism attraction or avoidance, and biocide release, among others.[299] Lethal entrainment of organisms in the intake pipes is one of the most direct impacts OTEC can have on the

292 Ibid., Article 79, para. 4.

293 Ibid., Articles 58, 87, 112 to 115.

294 Ibid., Article 79.

295 Ibid., Article 297.

296 Ibid., Article 79.

297 Ibid., para. 2.

298 Ibid., para. 3.

299 National Oceanic and Atmospheric Administration (NOAA), *Ocean thermal energy conversion: Assessing potential physical, chemical, and biological impacts and risks*, University of New Hampshire, 2010.

environment.[300]Also the vibration of the deep water pipe will create noise that may be detectable by marine mammals and fish, and acoustic monitoring of the site is necessary to be sure that the frequency emitted does not cause a large disturbance for marine mammal communication and navigation.[301] Biofouling can affect pipes and reduce the efficiency of the system.[302]

9 Ships, Including the People on Board

The establishment, operation and maintenance of installations and structures, cables and pipelines is dependent on logistical arrangements, including those provided by ships, as well as specialized workers and equipment. In particular, the growing market for offshore wind farms has created an industry requirement for construction and support vessels, which are also used to ferry workers and light cargoes to the wind turbines for construction and routine maintenance.[303] Vessels and equipment used for other offshore sectors may also be engaged in the construction of offshore wind farms.[304] For example, specialized cable installation vessels and underwater cable installation equipment are used to embed the electrical cables connecting the turbines together and the electrical grid.[305]

Article 94 of UNCLOS sets out some of the basic obligations of the flag State. It requires every State to effectively exercise its jurisdiction and control in administrative, technical and social matters over ships flying its flag. Flag States must take and enforce measures for ensuring safety at sea, *inter alia* with regard to the construction, equipment and seaworthiness of ships, the use of signals, the maintenance of communications and the prevention of collisions, that conform to generally accepted international regulations, procedures and

300 C.M. Comfort and L. Vega, 'Environmental Assessment of Ocean Thermal Energy Conversion in Hawaii' (2011), available at <http://hinmrec.hnei.hawaii.edu/wp-content/uploads/2010/01/Environmental-Assessment-of-OTEC-in-Hawaii1.pdf> accessed 7 September 2017, p. 5.

301 NOAA, n. 299, pp. 29–30.

302 K.V. Thomas and S. Brooks, 'The environmental fate and effects of antifouling paint biocides', *Biofoulin*, vol. 26(1), 2010, pp. 73–88.

303 Lloyd's Register – Guidance Notes for the Classification of Wind Farm Service Vessels, available at <www.lr.org/en/RulesandRegulations/guidance-notes-wind-farm-service-vessels.aspx> accessed 7 September 2017.

304 A. Chircop and P.L'Esperance, n. 62, pp. 439–487.

305 Transport Research Board, *Worker Health and Safety on Offshore Wind Farms – Special Report 310*, The National Academies Press, 2013, p. 29.

practices, as well as the manning of ships, labour conditions and the training of crews, taking into account the applicable international instruments. A State must cause an inquiry to be held by or before a suitably qualified person or persons into every marine casualty or incident of navigation on the high seas and the EEZ[306] involving a ship flying its flag and causing loss of life or serious injury to nationals of another State, or serious damage to ships or installations of another State or to the marine environment. The flag State and the other State must cooperate in the conduct of such inquiry.

UNCLOS also requires States, acting through the competent international organization or a general diplomatic conference, to establish international rules and standards to prevent, reduce and control pollution of the marine environment from ships and to re-examine them from time to time as necessary.[307] For the flag State, such global rules and standards constitute the minimum standards which it must adopt for vessels flying its flag[308] and enforce.[309] Coastal States and archipelagic States can adopt stricter rules and standards than the generally accepted global standards for application in their territorial sea, so long as such standards do not apply to the design, construction, manning or equipment of foreign ships or hamper innocent passage.[310] States bordering straits used for international navigation and archipelagic States may adopt laws and regulations relating to transit passage and archipelagic sea lanes passage in respect of the prevention, reduction and control of pollution, by giving effect to applicable international regulations regarding the discharge of oil, oily wastes and other noxious substances in the strait[311] and in the archipelagic sea lane.[312] Ships in transit passage and in archipelagic sea lanes passage must comply with generally accepted international regulations, procedures and practices for the prevention, reduction and control of pollution from ships,[313] amongst others.

Labour standards are contained in a number of conventions adopted by the International Labour Organization (ILO). The most important ILO Convention for seafarers is the 2006 Maritime Labour Convention, as amended.[314] It sets out,

306 UNCLOS, n. 10, Article 94, para. 7, and Article 58, para. 2.

307 Ibid., Article 211, para 1.

308 Ibid., para 2.

309 Ibid., Article 217.

310 Ibid., Article 21, para. 2, Article 24 and Article 52.

311 Ibid., Article 42, para. 1(b).

312 Ibid., Article 54.

313 Ibid., Article 39, para. 2(b) and Article 54.

314 ILO, *Maritime Labour Convention, 2006, as amended*, available at <http://www.ilo.org/wcmsp5/groups/public/---ed_norm/---normes/documents/normativeinstrument/wcms_554767.pdf> accessed 7 September 2017.

in one place, seafarers' rights to decent working conditions and covers almost every aspect of their work and life on board ships. The Convention requires ships to carry a Maritime Labour Convention Certificate and Declaration of Maritime Labour Compliance. In 2008, Guidelines for flag and port State control officers carrying out inspections under the Maritime Labour Convention[315] were adopted by a tripartite experts meeting with nearly 300 participants.[316]

The following IMO conventions have been deemed, on account of their worldwide acceptance, to fulfil the general acceptance requirement referred to in Article 94 of UNCLOS: SOLAS;[317] the 1966 International Convention on Load Lines,[318] and its Protocol of 1988; the 1969 International Convention on Tonnage Measurement of Ships;[319] COLREGS; the 1978 International Convention on Standards of Training, Certification and Watchkeeping for Seafarers,[320] (STCW 1978), as amended; and the 1979 International Convention on Maritime Search and Rescue[321] (SAR 1979).[322] Another IMO convention with an impact on design and construction standards is MARPOL.

As regards the impacts of activities involving ships on the marine environment, international rules and standards for the prevention of pollution have been developed by the IMO and include notably MARPOL. The 1990 International Convention on Oil Pollution Preparedness, Response and Cooperation and its 2000 Protocol on Preparedness, Response and Cooperation to Pollution Incidents by Hazardous and Noxious Substances[323] apply in case of a pollution incident, for example an incident caused by a collision. The disposal of wastes is regulated at the global level by the London Convention, as periodically amended by decisions of its Contracting Parties and the 1996

315 ILO, *Guidelines for port State control officers carrying out inspections under the Maritime Labour Convention, 2006* (ILO, 2009), available at <http://www.ilo.org/wcmsp5/groups/public/---ed_norm/---normes/documents/publication/wcms_101787.pdf> accessed 7 September 2017.

316 Contribution of the International Labour Organization to the Report of the Secretary-General on oceans and the law of the sea to the seventieth session of the General Assembly, available at <http://www.un.org/depts/los/general_assembly/contributions_2015/ILO.pdf> accessed 7 September 2017.

317 SOLAS, n. 189.

318 International Convention on Load Lines, 1966, 640 UNTS 133.

319 International Convention on Tonnage Measurement of Ships, 1969, 1291 UNTS 3.

320 International Convention on Standards of Training, Certification and Watchkeeping for Seafarers, 1978, 1361 UNTS 75.

321 International Convention on Maritime Search and Rescue, 1979, 1405 UNTS 118.

322 IMO, n. 187, p. 15.

323 Protocol on Preparedness, Response and Co-operation to Pollution Incidents by Hazardous and Noxious Substances, 2000, 2007 ATS 41.

Protocol. Other relevant treaties include the 2001 International Convention on the Control of Harmful Anti-fouling Systems on Ships, the 2004 International Convention for the Control and Management of Ships' Ballast Water and Sediments,[324] as well as several treaties addressing the liability and compensation for damages in connection with the carriage of oil and other hazardous and noxious substances by sea.[325]

The two international conventions of most relevance to the design and construction of offshore industry vessels are SOLAS and the International Convention on Load Lines. Both treaties apply only to vessels trading internationally, but will in general be recognized by administrations for ships on non-international voyages.[326] SOLAS applies to commercial vessels of 500 gross tonnage or more.[327] It defines a passenger ship as a ship which carries more than 12 passengers.[328] The International Convention on Load Lines applies to commercial vessels with a load line length of 24m or more.[329] If ships are not covered by these conventions or other legal instruments, they are subject to national laws and regulations.

IMO has adopted several non-mandatory legal instruments relating to offshore industry vessels, including the 1997 Code of Safe Practice for the carriage of cargoes and persons by offshore supply vessels,[330] the 2000 International Code of Safety for High-Speed Craft (HSC Code),[331] the 2006 Guidelines for the design and construction of offshore supply vessels,[332] the 2008 Code of Safety

324 IMO, 'Adoption of the Final Act and any Instruments, Recommendations, and Resolutions resulting from the work of the Conference: International Convention for the Control and Management of Ships' Ballast Water and Sediments, 2004. Text adopted by the Conference' (16 February 2004) IMO Doc BWM/CONF/36.

325 A detailed presentation of the relevant IMO instruments is provided in IMO, n. 187, pp. 84–86.

326 Danish Maritime Authority, *Review of maritime and offshore regulations and standards for offshore wind. Summary report on North Sea regulation and standards*, 2015, p. 4. Available at <https://www.dma.dk/Documents/Publikationer/ReportOnNorthSeaRegulationAndStandards.pdf > accessed 7 September 2017.

327 SOLAS, n. 189, Chapter I, Part A, Regulation 3(a)(ii).

328 Ibid., Chapter I, Part A, Regulation 2(f).

329 International Convention on Load Lines, n. 318, Article 5, para. 1(b).

330 IMO Assembly Res A.863(20) (27 November 1997).

331 IMO Maritime Safety Committee Res MSC.97(73) (14 December 2000), IMO Doc MSC 73/21/Add.1, annex 5.

332 IMO Maritime Safety Committee Res MSC.235(82) (22 December 2006), IMO Doc MSC 82/24/Add.2, annex 29.

for Special Purpose Ships (SPS Code),[333] and the 2009 Code for the Construction and Equipment of Mobile Offshore Drilling Units (MODU Code).[334] In 2012, a proposal was presented to the then IMO Sub-Committee on Ship Design and Equipment (DE)[335] to develop a more suitable classification of offshore wind farm construction vessels (OWFCV) and offshore wind farm service crafts (OWFSC), and non-mandatory guidelines clarifying the application of existing IMO instruments to these types of vessels, as well as the status of the specialized personnel carried on board and appropriate training and certification provisions for that personnel.[336] The proposal drew attention to the reported difficulties caused by the lack of a clear definition for industrial personnel and the lack of legally binding international safety standards for the carriage of more than 12 industrial personnel on board ships in the existing instruments. In response, the DE Sub-Committee agreed that there is an urgent need to further consider the application of existing IMO instruments to OWFCV and OWFSC, including a gap analysis, and to further clarify the application of relevant requirements concerning industrial personnel transported by sea.[337] In 2013, the DE Sub-Committee mandated the development of draft guidelines for offshore service craft and offshore construction vessels used in windfarm service, and guidelines addressing the carriage of more than 12 industrial personnel on board vessels engaged on international voyages.[338] In 2016, the Maritime Safety Committee, having decided that there was an urgent need for adequate and harmonized safety standards to address the carriage of more than 12 industrial personnel on board vessels engaged on international voyages, in particular for the offshore renewable energy sector, considered draft recommendations and mandated the Sub-Committee on

333 IMO Maritime Safety Committee Res MSC.266(84) (6 June 2008) IMO Doc MSC 84/24/Add.2, annex 17.

334 IMO Assembly Res A.1023(26) (18 January 2010).

335 The Sub-Committee was renamed Sub-Committee on Ship Design and Construction (SDC).

336 IMO, 'Classification of offshore industry vessels and consideration of the need for a code for offshore construction support vessels. Towards appropriate international standards for offshore industry vessels for the safe and sustainable exploitation of renewable energy sources. Submitted by Germany, CESA and ICS' (22 November 2011) IMO Doc DE 56/12.

337 IMO, 'Report to the Maritime Safety Committee' (28 February 2012) IMO Doc 56/25, para. 12.7.

338 IMO, 'Report to the Maritime Safety Committee' (5 April 2013) IMO Doc DE 57/25, para. 12.7; IMO, 'Report to the Maritime Safety Committee' (11 February 2014) IMO Doc SDC 1/26, paras. 18.8 and 19.16; IMO, 'Report to the Maritime Safety Committee' (10 March 2015) IMO Doc SDC 2/25, para. 9.8; and IMO, 'Report to the Maritime Safety Committee' (23 February 2016) IMO Doc SDC 3/21, para. 11.9.

Ship Design and Construction (SDC) to develop a new chapter of SOLAS and a new draft code on the safe carriage of more than 12 industrial personnel on board vessels engaged on international voyages.[339] The Committee agreed to delete the item 'Classification of offshore industry vessels and a review of the need for a non-mandatory code for offshore construction support vessels' from the biennial agenda of the SDC Sub-Committee.[340] In this connection, the Committee noted the information provided by the Chairman of the SDC Sub-Committee, that the ongoing work on that issue would be considered when developing standards for the safe carriage of more than 12 industrial personnel on board vessels engaged on international voyages.[341]

The Interim Recommendations on the safe carriage of more than 12 industrial personnel on board vessels engaged on international voyages, adopted subsequently by IMO,[342] which governments are invited to apply pending the planned development of the new chapter of SOLAS and the draft new code, are aimed at addressing the safe and efficient transfer of technicians at sea, such as those working in the growing offshore alternative energy sector. The Interim Recommendations define industrial personnel as all persons who are transported or accommodated on board for the purpose of offshore industrial activities performed on board other vessels and/or other offshore facilities. Offshore industrial activities covered by the Interim Recommendations include the construction, maintenance, operation or servicing of offshore facilities related, but not limited to exploration, the renewable or hydrocarbon energy sectors, aquaculture, ocean mining or similar activities.

The recommendations provide that industrial personnel should not be considered as passengers within the meaning of SOLAS regulation I/2(e). Before boarding the ship, industrial personnel should receive safety training, meeting some of the standards in the Code to the STCW Convention,[343] or appropriate alternatives. They should also *inter alia* receive on board ship specific safety familiarization.

---

339 IMO, 'Report of the Maritime Safety Committee on its ninety-sixth session" (31 May 2016) IMO Doc MSC 96/25, paras. 7.10–7.11, 7.15.

340 Ibid., para. 7.15.

341 Ibid.

342 IMO, Maritime Safety Committee Res MSC.418(97) 'Interim Recommendations on the safe carriage of more than 12 industrial personnel on board vessels engaged on international voyages' (6 December 2016) IMO Doc MSC 97/22/Add.1, annex 11.

343 IMO, 'The Manila Amendments to the Seafarers' Training, Certification and Watchkeeping (STCW) Code' (3 August 2010) IMO Doc STCW/CONF.2/34, also available at <http://www.imo.org/en/OurWork/HumanElement/TrainingCertification/Documents/34.pdf> accessed 7 Sepember 2017.

The need to ensure coherence with the Maritime Labour Convention in the development of the new SOLAS chapter has been underscored by the ILO.[344] The Maritime Labour Convention defines a 'seafarer' to mean 'any person who is employed or engaged or works in any capacity on board a ship to which this Convention applies',[345] and defines a ship as 'a ship other than one which navigates exclusively inland waters or waters within, or closely adjacent to, sheltered waters or areas where port regulations apply'.[346] It also provides that: '[i]n the event of doubt as to whether any categories of persons are to be regarded as seafarers for the purpose of this Convention, the question shall be determined by the competent authority in each Member after consultation with the shipowners' and seafarers' organizations concerned with this question'.[347] The ILO has also noted that in order to ensure uniform implementation of the Maritime Labour Convention, the International Labour Conference adopted in 2006 a resolution concerning information on occupation groups which establishes a set of criteria that should be considered by competent authorities when resolving doubts about designating a particular category of persons working on board as a seafarer. These include the following: (i) the duration of the stay on board of the persons concerned; (ii) the frequency of periods of work spent on board; (iii) the location of the person's principal place of work; (iv) the purpose of the person's work on board; and (v) the protection that would normally be available to the persons concerned with regard to their labour and social conditions to ensure they are comparable to that provided for under the Convention.[348]

10 Maritime Security Threats

Preserving and enhancing maritime security in the oceans and seas is central to the creation of an enabling environment for the sustainable production of offshore renewable energy. Maritime security involves protection from direct

---

344 IMO, 'Report of the Maritime Safety Committee on its ninety-seventh session' (2 December 2016) IMO Doc MSC 97/22, para. 6.7; and (6 December 2016) IMO Doc MSC 97/22/Add.1, annex 27, pp. 6–7.

345 Maritime Labour Convention, no. 314, Article II, para. 1(f).

346 Ibid., Article II, para. 1(i).

347 Ibid., Article II, para. 3.

348 ILO, 'Resolution concerning information on occupational groups' in *Resolutions adopted by the International Labour Conference at its 94th (Maritime) Session* (Geneva, February 2006), available at <http://www.ilo.org/wcmsp5/groups/public/@ed_norm/@normes/documents/publication/wcms_088130.pdf> accessed 7 September 2017, pp. 4–6.

threats to the territorial integrity of a State, such as an armed attack, as well as security from crimes at sea, such as piracy, armed robbery against ships, and terrorist acts.[349] However, intentional and unlawful damage to the marine environment, including from illegal dumping and the discharge of pollutants from vessels, are some activities that can also threaten the interests of States, particularly coastal States.[350]

Flag States play a particularly important role in maritime security, as they are required to effectively exercise jurisdiction and control in administrative, technical and social matters over ships flying their flag.[351] Flag States are responsible for ensuring that their vessels act in conformity with applicable rules of international law, wherever such vessels may be located.

In the territorial sea, coastal States and archipelagic States exercise sovereignty, while other States enjoy the right of innocent passage. However, the passage of foreign ships can be prevented if it is prejudicial to the peace, good order and security of the coastal State and archipelagic State. The rights, duties and jurisdiction of coastal States in the territorial sea, including in relation to innocent passage and the exercise of criminal jurisdiction on board a foreign ship, are contained in Part II of UNCLOS, section 3 thereof is also applicable to the right of innocent passage in archipelagic waters. In the contiguous zone, a coastal State can exercise the control necessary to prevent and punish infringement of its customs, fiscal, immigration or sanitary laws and regulations within its territory or territorial sea.[352] The legal regime governing straits used for international navigation is set out in Part III of UNCLOS. Certain provisions also apply, *mutatis mutandis*, to archipelagic sea lanes passage. UNCLOS provides for the enforcement of laws and regulations of the coastal State in the EEZ, including with respect to the protection and preservation of the marine environment.

Ships of all States enjoy the freedoms of navigation and overflight in the EEZ and on the high seas, as well as other internationally lawful uses of the sea related to these freedoms and compatible with the other provisions of the Convention. Part VII of UNCLOS includes specific provisions concerning the repression of piracy, which is defined in Article 101. All States have a duty to cooperate to the fullest possible extent in its repression,[353] and are granted

349 UNGA 'Report of the Secretary-General. Oceans and the law of the sea' (10 March 2008), UN Doc A/63/63, para. 39.

350 Ibid.

351 UNCLOS, no. 10, Article 94.

352 Ibid., Article 33.

353 Ibid., Article 100.

universal jurisdiction on the high seas to seize pirate ships and aircraft, or a ship or aircraft taken by piracy and under the control of pirates, and arrest the persons and seize the property on board.[354] UNCLOS also provides for the right of visit[355] and the right of hot pursuit,[356] which are particularly important in the context of maritime security. UNCLOS provides that enforcement action against foreign vessels can only be exercised by warships, military aircraft and other ships or aircraft clearly marked and identifiable as being on government service and authorized to that effect. Article 58 (2) of UNCLOS applies some of the high seas provisions and other pertinent rules of international law to the EEZ in so far as they are not incompatible with the provision of UNCLOS relating to the EEZ.[357]

It is important to distinguish the crime of piracy from armed robbery against ships, which can occur within the internal waters and territorial sea of a coastal State. In cases of armed robbery against ships, primary responsibility for enforcement normally falls on the coastal State. Armed robbery against ships also constitutes an offence under the 1988 Convention for the Suppression of Unlawful Acts against the Safety of Maritime Navigation (SUA Convention)[358] and, in some cases, the 2000 United Nations Convention against Transnational Organized Crime.[359]

Recognizing that shipping, offshore installations and other maritime interests could be potential targets for terrorist attacks, the IMO amended SOLAS in 2002 to introduce special measures to enhance maritime security on board ships.[360] It also amended the SUA Convention and the 1988 Protocol for the Suppression of Unlawful Acts against the Safety of Fixed Platforms Located on the Continental Shelf,[361] through the adoption of two Protocols. The 2005

354 Ibid., Article 105.

355 Ibid., Article 110.

356 Ibid., Article 111.

357 For further information, see UNGA, Report of the Secretary-General. Oceans and the law of the sea, n. 349.

358 Convention for the Suppression of Unlawful Acts against the Safety of Maritime Navigation, 1988, 1678 UNTS 222.

359 United Nations Convention against Transnational Organized Crime, 2000, 2225 UNTS 209.

360 IMO, Amendments to the 1974 SOLAS Convention and adoption of the International Code for the Security of Ships and of Port Facilities (12 December 2002) IMO Docs SOLAS/CONF.5/32; and (17 December 2002) SOLAS/CONF.5/34, annex 1.

361 Protocol for the Suppression of Unlawful Acts against the Safety of Fixed Platforms Located on the Continental Shelf, 1988, 1678 UNTS 304.

Protocol to the SUA Convention[362] has the effect of broadening the list of offences to include, for example, the offences of using a ship itself in a manner that causes death or serious injury or damage, transporting a biological, chemical or nuclear (BCN) weapon, knowing it to be such, and transporting any equipment, materials or software or related technology that significantly contributes to the design, manufacture or delivery of a BCN weapon, with the intention that it be used for such purpose. It is also an offence to unlawfully and intentionally transport a person on board a ship knowing that the person has committed an offence under the SUA Convention or an offence set forth in any of the conventions listed in the Annex to the SUA Convention. The 2005 Protocol introduces provisions for the boarding of a ship where there are reasonable grounds to suspect that the ship or a person on board the ship is, has been, or is about to be involved in the commission of an offence under the SUA Convention. The Preamble to the 2005 Protocol to the SUA Convention recalls the importance of UNCLOS and of the customary international law of the sea.[363]

Action against persons committing unlawful acts against fixed platforms on the continental shelf can be taken on the basis of the 1988 Protocol for the Suppression of Unlawful Acts against the Safety of Fixed Platforms Located on the Continental Shelf and its 2005 Protocol[364] which broadened the list of offences, to include *inter alia* the placement or causing the placement on a fixed platform, by any means whatsoever, of a device or substance which is likely to destroy that fixed platform or likely to endanger its safety.[365] The Protocol only applies to fixed platforms which Article 1 of the 1988 Protocol defines to mean an artificial island, installation or structure permanently attached to the sea-bed for the purpose of exploration or exploitation of resources or for other economic purposes.

---

362 IMO, 'Adoption of the Final Act and any Instruments, Recommendations and Resolutions resulting from the Work of the Conference. Protocol of 2005 to the Convention for the Suppression of Unlawful Acts against the Safety of Maritime Navigation. Text adopted by the Conference' (1 November 2005) IMO Doc LEG/CONF.15/21.

363 IMO, n. 187, p. 53.

364 IMO, 'Adoption of the Final Act and any Instruments, Recommendations and Resolutions resulting from the Work of the Conference. Protocol of 2005 to the Protocol for the Suppression of Unlawful Acts against the Safety of Fixed Platforms Located on the Continental Shelf. Text adopted by the Conference' (1 November 2005) IMO Doc LEG/CONF.15/22.

365 Ibid., Article 3.

11 Compliance, Enforcement, Responsibility and Liability and Settlement of Disputes

Many of the activities relating to offshore renewable energy are conducted by private operators, for example the owners of the infrastructure or vessels carrying out the energy activities, or their charterers. It is therefore also important to consider the rules that have been developed by and for the industry.[366] For example, standards of conduct that have been specifically tailored to offshore renewable energy by specialized professional organizations include those of the International Electrotechnical Commission.[367]

Particularly noteworthy are also the general principles of responsible conduct developed for the industry by the UN Global Compact and the Organization for Economic Cooperation and Development (OECD). The ten principles of the UN Global Compact include recommendations that businesses should support a precautionary approach to environmental challenges, undertake initiatives to promote greater environmental responsibility and encourage the development and diffusion of environmentally friendly technologies.[368] The OECD Guidelines for Multinational Corporations recommend *inter alia* that enterprises should, within the framework of laws, regulations and administrative practices in the countries in which they operate, and in consideration of relevant international agreements, principles, objectives, and standards, take due account of the need to protect the environment, public health and safety, and generally to conduct their activities in a manner contributing to the wider goal of sustainable development.[369]

---

366 A. Bonfanti and F. Romanin Jacur, 'Energy from the Sea and the Protection of the Marine Environment: Treaty-Based Regimes and Ocean Corporate Social Responsibility', *The International Journal of Marine and Coastal Law*, vol. 29(4), 2014, pp. 622–644.

367 Ibid.; IRENA, *International Standardization in the Field of Renewable Energy* (IRENA, 2013) 1–80, available at <http://www.irena.org/DocumentDownloads/Publications/International_Standardisation_%20in_the_Field_of_Renewable_Energy.pdf> accessed 7 September 2017, and inventory of standards relevant to renewable energy used for the International Standardization in the Field of Renewable Energy, available at <http://www.irena.org/DocumentDownloads/Publications/Inventory_renewable_energy_standards.pdf> accessed 7 September 2017.

368 See principles 7–9 in UN, *The Ten Principles of the UN Global Compact*, available at <https://www.unglobalcompact.org/what-is-gc/mission/principles> accessed 7 September 2017.

369 OECD, 'OECD Guidelines for Multinational Enterprises' (2011), available at <http://www.oecd.org/daf/inv/mne/48004323.pdf> accessed 7 September 2017, pp. 42–46.

UNCLOS places the primary responsibility for implementation and enforcement of its obligations on States. Those provisions that are particularly relevant for the sustainable production of offshore renewable energy are presented below and should be considered with those of the previous section on how to prevent and address threats to maritime security.

For example, States are required to ensure compliance by ships flying their flag or of their registry, which in many treaties are defined to also include fixed and floating platforms, with applicable international rules and standards, established through the competent international organization or general diplomatic conference, and with their laws and regulations adopted in accordance with UNCLOS for the prevention, reduction and control of pollution of the marine environment from ships.[370]

States are also required to enforce their laws and regulations and take other measures necessary to implement applicable international rules and standards established through competent international organizations or diplomatic conferences to prevent, reduce and control pollution arising from or in connection with seabed activities subject to their jurisdiction and from artificial islands, installations and structures under their jurisdiction, pursuant to Articles 60 and 80.[371]

In cases of non-compliance, the coastal State has the right under UNCLOS to take the necessary steps in its territorial sea to prevent passage that is not innocent,[372] for example if a foreign ship engages in any act aimed at interfering with any facilities or installations of the coastal State. It may also exercise criminal jurisdiction on board a foreign ship passing through the territorial sea to arrest any person or to conduct any investigation in connection with crimes committed on board the ship during its passage in certain specified circumstances set out in UNCLOS,[373] including if the consequences of the crime extend to the coastal State, or if the crime is of a kind to disturb the peace of the country or the good order of the territorial sea. It may also exercise such criminal jurisdiction if the ship violated its laws and regulations in the EEZ adopted in accordance with the Convention, or in order to exercise its port or coastal State enforcement jurisdiction as provided for in Part XII.

UNCLOS further grants the coastal State the right of hot pursuit against a foreign ship in its internal waters, archipelagic waters, territorial sea or

---

370 UNCLOS, n. 10, Article 217, see also section 9 above on 'Ships, including the People on Board'.

371 Ibid., Article 214.

372 Ibid., Article 25.

373 Ibid., Article 27.

the contiguous zone in situations when the coastal State has good reason to believe that the ship has violated its laws and regulations. The right of hot pursuit applies *mutatis mutandis* to violations in the EEZ and on the continental shelf, including in relation to safety zones around continental shelf installations, of the laws and regulations of the coastal State.[374] In the Area, the International Seabed Authority has the right to *inter alia* inspect all installations used in connection with activities in the Area[375] and also prevent interference with other activities in the marine environment.[376]

11.1 *Responsibility and Liability*

Article 235 of UNCLOS provides that States are responsible for the fulfilment of their international obligations concerning the protection and preservation of the marine environment and that they are liable in accordance with international law. They are required to ensure that resource is available in accordance with their legal systems for prompt and adequate compensation or other relief in respect of damage caused by pollution of the marine environment by natural or juridical persons under their jurisdiction. The Convention emphasizes the need to develop norms of international law dealing with responsibility and liability for the assessment and compensation for damage and the settlement of related disputes, as well as where appropriate, to develop criteria and procedures for payment of adequate compensation, such as compulsory insurance or compensation funds. Article 304 of UNCLOS provides that its provisions are without prejudice to the application of existing rules and the development of further rules regarding responsibility and liability under international law.

The international law on responsibility and liability has been further developed by the International Law Commission. It adopted Draft Articles on the Responsibility of States for Internationally Wrongful Acts in 2001,[377] as well as Draft Articles on the Prevention of Transboundary Harm from Hazardous Activities.[378] In 2006, the International Law Commission adopted the text of the preamble and a set of eight draft principles on the allocation of loss in the case of transboundary harm arising out of hazardous activities with

---

374 Ibid., Article 111.

375 Ibid., Article 153(5).

376 Ibid., Annex III, Article 17(1) (ix).

377 ILC, 'Report of the International Law Commission on the Work at its 53rd session' (23 April–1 June and 2 July-10 August 2001), pp. 26–30. For the text of the draft articles with commentaries, see pp. 31–143.

378 Ibid., pp. 146–148. For the text of the draft articles with commentaries, see pp. 148–170.

commentaries.[379] The preamble reaffirms Principles 13 and 16 of the Rio Declaration on Environment and Development.[380] The need to develop national law regarding liability and compensation for the victims of pollution and other environmental damage is stressed in Principle 13. Principle 16 addresses the promotion of internalization of environmental costs, taking into account the polluter-pays principle.[381]

Treaties that have been specifically developed in the field of civil liability and compensation for damage to the marine environment are mainly limited to the areas of maritime transport of oil, as well as the carriage of hazardous and noxious substances by sea.

The advisory opinions of the International Tribunal for the Law of the Sea have clarified the extent of responsibility and liability of States and sponsored persons and entities undertaking activities in the Area and those of flag States on the high seas.[382] In its *Advisory Opinion on the Responsibilities and obligations of States sponsoring persons and entities with respect to activities in the Area* (hereinafter *Case No. 17*), the Seabed Disputes Chamber of the Tribunal recognized the distinct liability of a sponsored contractor for the failure to comply with its obligations and the liability of sponsoring States for not complying with their own set of obligations.[383] Having noted that, in accordance with Article 139 of UNCLOS and Article 4 of Annex III thereto, the existence of damage is essential for the establishment of the liability of the State,[384] the Chamber ruled that 'the liability of sponsoring States arises from their failure to carry out their own responsibilities and is triggered by the damage caused by sponsored contractors. There must be a causal link between the sponsoring State's failure and the damage, and such a link cannot be presumed'.[385] Where the sponsoring State has met its obligations, damage caused by the sponsored

379 ILC, 'Report of the International Law Commission on the Work of its 58th session' (1 May-9 June and 3 July–11 August 2006), pp. 105–182, includes the commentaries.

380 'Report of the United Nations Conference on Environment and Development' (Rio de Janeiro, 3–14 June 1992) UN Doc A/CONF.15/26 (Vol. I).

381 ILC, n. 379, p. 115.

382 *Responsibilities and obligations of States sponsoring persons and entities with respect to activities in the Area*, Advisory Opinion, ITLOS Reports 2011, p. 10; *Request for an advisory opinion submitted by the Sub-Regional Fisheries Commission (SRFC)*, Advisory Opinion, ITLOS Reports 2015, p. 4.

383 *Responsibilities and obligations of States with respect to activities in the Area*, n. 382, p. 57, paras. 171–172.

384 Ibid., p. 58, para. 178.

385 Ibid., p. 60, para. 184.

contractor does not give rise to the sponsoring State's liability.[386] If the sponsoring State has failed to fulfil its obligations but no damage has occurred, the consequences of such wrongful act are determined by customary international law.[387] The Seabed Disputes Chamber stated that '[t]he sponsoring State's obligation "to ensure" is not an obligation to achieve, in each and every case, the result that the sponsored contractor complies with the obligations. Rather, it is an obligation to deploy adequate means, to exercise best possible efforts, to do the utmost, to obtain this result.... [It] may be characterized as an obligation "of conduct" and not "of result", and as an obligation of "due diligence".'[388] The Chamber clarified the obligation of 'due diligence' requires the sponsoring State to adopt laws and regulations and to take administrative measures, which are within the framework of its legal system 'reasonably appropriate' for securing compliance by persons under its jurisdiction. The standard of 'due diligence' may vary over time depending on the level of risk and on the activities involved. Turning to the second 'due diligence' obligation, the Seabed Disputes Chamber described it a direct obligation for the sponsoring State with which it must comply independently of its obligation to ensure a certain conduct on the part of the sponsored contractors.

Similar to the approach of the Seabed Disputes Chamber in *Case No. 17*, the Tribunal held in the *Request for Advisory Opinion submitted by the Sub-Regional Fisheries Commission*, that the liability of a flag State arises only if it fails to comply with its 'due diligence' obligations.[389] The flag State is not liable if it has taken all necessary and appropriate measures to meet its 'due diligence' obligations.[390]

11.2 *Settlement of Disputes*

UNCLOS sets out a comprehensive legal regime for the settlement of disputes between States Parties concerning the interpretation or application of UNCLOS in its Part XV and Annexes V to VIII. States Parties are required to settle any disputes concerning the interpretation or application of the Convention by peaceful means.[391] Section 2 of Part XV of UNCLOS sets out compulsory procedures entailing binding decisions for the resolution of disputes. Disputes

386 Ibid., p. 76, operative para. 4.

387 Ibid.

388 Ibid., p. 41, para. 110.

389 *Request for an advisory opinion submitted by the Sub-Regional Fisheries Commission (SRFC)*, n. 382, p. 44, para. 146.

390 Ibid., p. 45, para. 148.

391 UNCLOS, n. 10, Article 279.

concerning the interpretation or application of UNCLOS and the interpretation or application of an international agreement related to the purposes of the Convention can be submitted to the International Tribunal for the Law of the Sea, the International Court of Justice, an arbitral tribunal constituted in accordance with Annex VII, or a special arbitral tribunal constituted in accordance with Annex VIII to the Convention.[392] Conciliation is also available and, in certain circumstances, submission to it would be compulsory.[393]

12 Cooperation and Coordination

Offshore renewable energy requires an extensive spatial deployment for the investment to be commercially sufficient,[394] which has implications for other uses of the seas. Reconciling potential conflicts with other uses of marine space, such as navigation, fishing, aquaculture, tourism, oil and gas exploration and exploitation, laying of cables and pipelines, overflight, seabed mining, military exercises, nature development and conservation, underwater cultural heritage protection,[395] land reclamation, and recreation, will thus be vitally important if the production of offshore renewable energy is going to be able to reach its full potential.[396] Furthermore, in order to achieve sustainable production of offshore renewable energy, activities have to be managed in a manner that ensures that the three pillars of sustainable development are sufficiently integrated and balanced.

Uses of the oceans can only be effectively balanced and integrated from a sustainable development perspective if a supporting policy framework exists, which fosters integrated approaches to the management of human activities in the oceans and seas; the establishment of supporting institutional mechanisms for cross-sectoral cooperation and coordination and effective stakeholder participation.

UNCLOS recognizes in its preamble that the problems of ocean space are closely interrelated and need to be considered as a whole. It sets out a carefully balanced and equitable package of rights and duties in the various maritime

---

392 Ibid., Article 287 and 288.

393 Ibid., Articles 284, 297, para. 2(b) and para. 3(b), 298, para. 1(a), and Annex V.

394 M. Gavouneli, n. 84.

395 UNESCO, 'Developing the seabed: resource extraction and energy development at sea', available at <http://www.unesco.org/new/en/culture/themes/underwater-cultural-heritage/protection/threats/developing-the-seabed> accessed 7 September 2017.

396 D. Leary and M. Esteban, n. 247, p. 646.

zones where the enjoyment of rights and benefits entails the concomitant undertaking of duties and obligations. If effectively implemented together with the legal instruments that complement it, the Convention can provide an enabling legal environment for the sustainable production of offshore renewable energy.

The United Nations General Assembly has consistently recognized the need to improve cooperation and coordination at the national, regional and global levels, in accordance with UNCLOS, to support and supplement, *inter alia*, the efforts of each State in promoting the implementation and observance of the Convention and the integrated management and sustainable development of the oceans and seas.[397]

In the discussions on marine renewable energies at the thirteenth meeting of the UN Open-ended Informal Consultative Process on Oceans and the Law of the Sea, delegations drew attention to marine spatial planning as an important tool to sustain ecosystem functions and services, provide for public access and increase certainty and predictability for economic investments. Other tools identified at the meeting included strategic environmental assessments and ecosystem-based management.[398]

Marine spatial planning and strategic environmental assessments can also constitute valuable tools for supporting streamlined approval processes for offshore renewable energy projects which are essential for securing investments. Indeed, simplification of the regulatory regime for offshore renewable energy provides more clarity for project developers and confidence for potential investors.[399] The channeling of responsibility for the administration of such energy through a central agency offers the opportunity for a more efficient system through the reduction of conflict and alignment of strategic policy objectives.[400]

While several States have adopted national targets for offshore renewable energy, many have not yet adopted specific legislation and tend to use existing regulatory requirements applicable to other sectors (for example oil and

397 UNGA Res 71/257, n. 7, preambular para. 13.

398 UNGA 'Report on the work of the United Nations Open-ended Informal Consultative Process on Oceans and the Law of the Sea at its thirteenth meeting', n. 8, para. 57.

399 National Energy Administration of China and the World Bank, *China: Meeting the Challenges of Offshore and Large-Scale Wind Power*, The World Bank Asia Sustainable and Alternative Energy Program, 2010, p. 35, available at <http://siteresources.worldbank.org/EXTEAPASTAE/Resources/ASTAE-China-Offshore-wind-web.pdf> accessed 7 September 2017.

400 Ibid.

gas) which may not be suitable and may lead to delays in the approval process. Some States have streamlined the approval process by establishing a comprehensive approach to siting and approving ocean energy projects through a combination of a simplified agency structure – a 'one stop shop' approach – and marine spatial planning.[401] However, many States have not yet applied the management tools that exist to offshore renewable energy. For example, it has been observed that while many European Union member States are implementing sophisticated planning systems governing the use of maritime space in areas within national jurisdiction, spatial planning procedures followed by the majority of those States are not targeted specifically at offshore renewable energy development or at resolving conflict between the construction and operation of wind farms and other maritime activities.[402]

Conflicting regulatory policies can be avoided by establishing clear and streamlined regulatory frameworks at the national level, including national legislation that takes a cross-sectoral approach and is not limited to the territorial sea.[403]

Cooperation among States is also important and beneficial. Many renewable energy installations and structures are, like submarine cables, enterprises comprised of single and multi-national companies and create an opportunity for enhanced cooperation and coordination between States. The benefits of such approach are demonstrated by the North Sea Countries Offshore Grid Initiative.[404] The Political Declaration on energy cooperation between the North Seas Countries[405] provides a framework for trading of energy, data sharing, coordination of permitting procedures, and harmonization of technical rules, amongst other objectives.

Another recent example of regional cooperation includes the Martinique Action Plan for Renewable Energy Development on Islands, adopted at the

---

401 M.E. Higgins and J. Busch, 'Offshore Wind and Wave Energy and Ocean Governance' in R.S. Abate (ed.), *Climate Change Impacts on Ocean and Coastal Law: US and International Perspectives*, Oxford University Press, 2015, p. 183.

402 R. Long, 'Harnessing Offshore Wind Energy: Legal Challenges and Policy Conundrums in the European Union', *International Journal of Marine and Coastal Law*, vol. 29(4), p. 707.

403 D. Leary and M. Esteban, n. 247, pp. 646–647.

404 *Memorandum of Understanding on the North Seas Countries Offshore Grid Initiative*, no. 165.

405 *Political Declaration on energy cooperation between the North Seas Countries and action plan*, n. 167.

Martinique Conference on Island Energy Transitions, held from 22 to 24 June 2015.[406]

At the thirteenth meeting of the UN Open-ended Informal Consultative Process on Oceans and the Law of the Sea, many delegations also emphasized the need to strengthen international cooperation to address challenges in the development of marine renewable energies in accordance with the current international legal framework.[407] Capacity-building as well as the transfer of technology were considered essential elements of international cooperation and coordination and Part XIV of UNCLOS on development and transfer of marine technology was highlighted.[408] Many delegations emphasized that sustainable funding, technology transfer and capacity-building were critical to ensuring that developing countries with potential offshore renewable energy, in particular small island developing States, could benefit therefrom.[409] Some delegations underscored the need for further efforts in this regard, since marine renewable energies were still at the early stages of development and were expensive to deploy.[410] In addition, most of the resources needed to generate marine renewable energies were not accessible to developing countries.[411]

13 Conclusions

Effective regulatory and policy frameworks, an enabling environment and legal certainty are critical for scaling up investments for technological development for the sustainable production of offshore renewable energy.

International law provides a regulatory framework that is broadly suitable for the development of renewable energy.[412] However, there are limitations. The regulation of offshore renewable energy is not covered by a single instrument; its regulation is dispersed not only across various areas of international

406 *Martinique Action Plan For Renewable Energy Development On Islands*, available at <http://irena.org/eventdocs/irena%20-%20map%2010%20july%20final.pdf> accessed 7 September 2017.

407 UNGA 'Report on the work of the United Nations Open-ended Informal Consultative Process on Oceans and the Law of the Sea at its thirteenth meeting', n. 8, para. 9.

408 Ibid.

409 Ibid., para. 50.

410 Ibid.

411 Ibid.

412 M.A. Costelos, n. 51, p. 232.

law, but also in a variety of national regulations.[413] In particular, the overall international governance framework that has evolved to address the risks posed by offshore renewable energy projects to the marine environment is largely fragmented and recommendatory in nature.[414] Coastal States are left broad discretion to develop offshore renewable energy projects subject only to general law of the sea responsibilities and requirements, such as the conduct of environmental impact assessments.[415]

UNCLOS, as the core legal instrument, can be further built upon and complemented by other instruments and can also provide a unifying framework for the international regulation of offshore renewable energy, thereby also promoting legal certainty for technological development and investment. However, the rights and duties of States with respect to some of the technologies require clarification and also need to take into account the rapid advances in technological developments.

Effective implementation of the legal regime in a coordinated and integrated manner is also critical in order to avoid fragmentation[416] and accommodate and balance the diverse uses of the ocean from the perspective of the three pillars of sustainable development. Management tools, such as marine spatial planning, can support those objectives. There is also a need to foster cooperation and coordination among States and coherence of the legal framework in view of the transboundary nature of some of the technologies involved, which a predominantly self-regulated industry may not be able to foster nor ensure that the three pillars of sustainable development are sufficiently integrated and balanced. Suggestions that have been put forward include the development of a legal instrument which would afford governments and industry the opportunity to clarify their respective obligations and address particular interests of mutual benefit.[417] There is clearly a need for an institutional framework or governance mechanism at the global level that has

413 Ibid., p. 228.

414 S. Mc Donald and D. L. VanderZwaag, n. 51, p. 325.

415 Ibid., p. 326.

416 T. Scovazzi and I. Tani, 'Offshore Wind Energy Development in International Law' in J. Ebbesson et al. (eds.), *International Law and Changing Perceptions of Security*, Brill, 2014, p. 258.

417 Suggestions include the development of a protocol, see M. Tsamenyi and M. Herriman, 'Ocean Energy and the Law of the Sea: The Need for a Protocol', *Ocean Development and International Law*, vol. 29, 1998, pp. 3–19, and the development of a code of best practices with regard to the laying and repair of submarine cables and the conduct of cable-routing surveys, see Center for International Law (CIL) 'Workshop report on submarine cables and law of the sea: 14–15 December 2009, Singapore', available at <https://cil.nus

or can take responsibility to consider the issue from an integrated and holistic perspective and in a systematic manner,[418] since there is no single international organization with a mandate covering offshore renewable energy.[419] An institutional framework or governance mechanism could also play an important role in promoting international cooperation between States, including in relation to capacity-building and the transfer of technology for developing States, in order to promote the development of offshore renewable energy.

The High-level Political Forum, which will be reviewing the implementation of the 2030 Agenda for Sustainable Development, including Sustainable Development Goal 7, is not expected to focus specifically on offshore renewable energy. The United Nations General Assembly is the only global policy-making body that has thus far considered, with the assistance of its subsidiary body, the UN Open-ended Informal Consultative Process on Oceans and the Law of the Sea, offshore renewable energy from a comprehensive, cross-sectoral and integrated perspective. As the global institution having the competence to undertake an annual consideration and review of the overall developments relating to ocean affairs and the law of the sea,[420] the General Assembly is well placed to revisit the issue and recommend a course of action that would ensure that the legal regime at the global level relating to offshore renewable energy provides the necessary legal certainty for technological development and investment and the sustainable production of offshore renewable energy, and would enable the accommodation of the various uses in the maritime zones in a balanced manner, as well as provide an opportunity to enhance international cooperation between States, including in relation to capacity-building and the transfer of marine technology for developing countries.

Bibliography

Appiott, J., Cicin-Sain, B. & Dhanju, A., 'Encouraging renewable energy in the offshore environment', *Ocean and Coastal Management*, vol. 90, 2014, pp. 58–64.

Binger, A., *Potential and Future Prospects for Ocean Thermal Energy Conversion (OTEC) In Small Island Developing States (SIDS)*, 2004.

.edu.sg/wp/wp-content/uploads/2009/10/Workshop-Report-29-Jan-2010.pdf> accessed 7 September 2017.

418 M.A. Costelos, n. 51, p. 232.

419 S. Mc Donald and D.L. VanderZwaag, n. 51, p. 325.

420 UNGA Res. 49/28 'Law of the Sea' (19 December 1994); UNGA Res. 71/257, n. 7, para. 352.

Bonfanti, A. and Romanin Jacur, F., 'Energy from the Sea and the Protection of the Marine Environment: Treaty-Based Regimes and Ocean Corporate Social Responsibility', *The International Journal of Marine and Coastal Law*, vol. 29, no. 4, 2014, pp. 622–644.

Brooks, S. and Thomas, K.V., 'The environmental fate and effects of antifouling paint biocides', *Biofouling*, vol. 26, no. 1, 2010, pp.73–88.

Busch, J. and Higgins, M.E., 'Offshore Wind and Wave Energy and Ocean Governance', *Climate Change Impacts on Ocean and Coastal Law: US and International Perspectives*, Oxford University Press, 2015, pp. 153–192.

Castelos, M.A., 'Marine Renewable Energies: Opportunities, Law and Management', *Ocean Development and International Law*, vol. 45, 2014, pp. 221–237.

Chircop, A. and L'Esperance, P., 'Functional Interactions and Maritime Regulation: The Mutual Accommodation of Offshore Wind Farms and International Navigation and Shipping', *Ocean Yearbook*, vol. 30, 2016, pp. 439–487.

Churchill, R.R. and Lowe, A.V., *The law of the sea*, Manchester University Press, 1999.

Comfort, C.M. and Vega, L., *Environmental Assessment of Ocean Thermal Energy Conversion in Hawaii*, University of Hawaii, 2011.

Davenport, T. and Nielsen, W., 'Submarine Cables and Offshore Energy' in D. Burnett, R. Beckman, and T. Davenport (eds.), *Submarine Cables: The Handbook of Law and Policy*, Brill, 2014, pp. 351–373.

Esteban, M. and Leary, D., 'Climate Change and Renewable Energy from the Ocean and Tides: Calming the Sea of Regulatory Uncertainty', *International Journal of Marine and Coastal Law*, vol. 24, 2009, pp. 617–651.

Fink, C. *The International Regulation of Offshore Wind Farms: under the 1982 Law of the Sea Convention (UNCLOS)*, University of Oslo, 2005.

Galea, F., 'A Legal Regime for the Exploration and Exploitation of Offshore Renewable Energy', *Oceans Yearbook*, vol. 25, 2011, pp. 101–129.

Gavouneli, M., 'Energy Installations in the Marine Environment' in J. Barrett and R. Barnes (eds.), *Law of the Sea: UNCLOS as a living treaty*, British Institute of International and Comparative Law, 2016, pp. 187–208.

Herriman, M. and Tsamenyi, M., 'Ocean Energy and the Law of the Sea: The Need for a Protocol', *Ocean Development and International Law*, vol. 29, 1998, pp. 3–19.

IRENA, *Innovation Outlook, Offshore Wind, Summary for Policy Makers*, 2016.

IRENA, *Renewable Energy Capacity Statistics*, 2017.

IRENA, *Innovation Outlook: Offshore Wind*, 2016.

IRENA, *Renewable Energy and Jobs*, 2013.

Joseph, S.L., 'Legal Issues Confronting the Exploitation of Renewable Sources of Energy from the Oceans', *California Western International Law Journal*, vol. 11, 1981, pp. 387–424.

Kindt, J.W., 'Ocean Thermal Energy Conversion', *Georgia Journal of International and Comparative Law*, vol. 14, no. 1, 1984, pp. 1–27.

Long, R., 'Harnessing Offshore Wind Energy: Legal Challenges and Policy Conundrums in the European Union', *International Journal of Marine and Coastal Law*, vol. 29, no. 4, 2014, pp. 690–715.

Lund, N.J., 'Renewable Energy as a Catalyst for Changes to the High Seas Regime', *Ocean and Coastal Law Journal*, vol. 15, 2010, pp. 95–125.

McCarthy, E., *International Regulation of Underwater Sound: Establishing Rules and Standards to Address Ocean Noise Pollution*, Kluwer, 2004

McDonald, S. and VanderZwaag, D.L., 'Renewable Ocean Energy and the International Law and Policy Seascape: Global Currents, Regional Surges', *Ocean Yearbook*, vol. 29, 2011, pp. 299–326.

Müller, H.K. and Roggenkamp, M., 'Regulating offshore energy sources in the North Sea – Reinventing the wheel or a need for more coordination?', *The International Journal of Marine and Coastal Law*, vol. 29, no. 4, 2014, pp. 716–737.

Nanda, V.P., 'The Legal Framework for the Development of Ocean Thermal Energy Conversion (OTEC)', *San Diego Law Review*, vol. 19, 1982, pp. 385–408.

Nandan, S.N. and Rosenne, S., *United Nations Convention on the Law of the Sea: A Commentary*, vol. II, Brill/Nijhoff, 1993.

Nandan, S.N. and Rosenne, S., *United Nations Convention on the Law of the Sea: A Commentary*, vol. III, Brill/Nijhoff, 1995.

Nielsen, W. and Davenport, T., 'Submarine Cables and Offshore Energy' in D. Burnett, R. Beckman, T. Davenport (eds), *Submarine Cables: The Handbook of Law and Policy*, Brill, 2014, pp. 352–373.

'Report of the IWC Scientific Committee Workshop on Interactions between Marine Renewable Projects and Cetaceans Worldwide', IWC SC/64/Rep6 Rev1.

Robertson, A.N. and Jonkman, J.M., *'Load Analysis of Several Offshore Floating Wind Turbine Concepts'*, ISOPE, vol. 1, 2011, p. 443–450.

Roeben, V., 'Governing Shared Offshore Electricity Infrastructure in the Northern Seas' in *International and Comparative Law Quarterly*, vol. 62, no. 4, 2013, pp. 839–864.

Scott, K., 'Tilting at offshore windmills: Regulating wind farm development within the renewable energy zone', *Journal of Environmental Law*, vol. 18, no. 1, 2006, pp. 89–118.

Scovazzi, T. and Tani, I., 'Offshore Wind Energy Development in International Law' in J. Ebbesson et al. (eds.), *International Law and Changing Perceptions of Security*, Brill, 2014, pp. 244–258.

United Nations, *The First Global Integrated Marine Assessment: World Ocean Assessment I*, Cambridge University Press, 2017.

Wegelein, F.H. Th., *Marine Scientific Research. The Operation and Status of Research Vessels and other Platforms in International Law*, Nijhoff, 2005.

CHAPTER 3

Sustainable Energy Generation from the Oceans

Henning Jessen

1 Rising Global Energy Demand and the Impact on the (Offshore) 'Energy Mix'

It has been pointed out in the relevant academic literature that one of the main global future challenges for mankind will be to guarantee energy supply at a reasonable price.[1] This future challenge will also have to take into account the rising energy demand at the global scale. Transformations in the global energy system will result in renewable energy and natural gas (possibly even methane gas) to be the main energy sources to meet human energy demand growth until 2040.[2] The important advantages of renewables and the resulting political euphoria for harnessing the economic and ecologic potentials of wind power, tidal resources and solar energy tends to whitewash one important fact: Large parts of the 21st century will still fall into the era of fossil fuels and other ways of utilizing fossil energy resources – both onshore and offshore. The former dominance of coal and oil is replaced by a much more sophisticated 'energy mix' which, however, still includes a major share of non-renewable fossils. In some parts of the world the non-renewable sector will even continue to grow as new exploration and exploitation projects will be realized. In fact, projections until 2040 indicate that global oil consumption will rise moderately to 103.5 million barrels per day (mb/d) as compared to 92.5 mb/d in 2015 – even taking into account all voluntary climate change mitigation mechanisms to be implemented in accordance with the Paris Agreement of 2015.

As a result, both onshore and offshore, renewable energy production will not completely replace conventional forms of non-renewable energy generation. Rather, there will be a continuous process of modernizing and complementing the global 'energy mix'. This will have a significant effect on the oceans and also on the competition of coastal States and private stakeholders for ocean space. Effectively, energy-related human activities at sea will continue to multiply. The potential for mutual interferences of offshore activities with other

1 See only N. Banks and S. Trevisanut, *Energy from the Sea*, Brill, 2015, p. 1.

2 See International Energy Agency (IEA), 'World Energy Outlook 2016', <http://www.iea.org/newsroom/news/2016/november/world-energy-outlook-2016.html> accessed 1 July 2017.

 | DOI 10.1163/9789004360273_004

activities will rise. Ocean governance faces multiple challenges, above all, in relation to dangers of becoming too fragmented and incoherent in balancing different sustainability challenges. The variety of human activities – both energy-related and in relation to other economic areas such as tourism, fishing or telecommunication – will result in greater risks. These risks can manifest themselves as 'vertical risks' (i.e., the offshore oil and gas industry moving out towards greater water depths and more remote exploration targets) or as 'horizontal risks' (i.e., different ways of utilizing sea areas at the same level, such as offshore installations moving closer to established traffic lanes or electrical transmission lines and natural gas pipelines on the seabed potentially conflicting with seabed fishing or mining). Consequently, there is a growing need for transboundary marine spatial planning and for integrating the planning activities with energy and climate policy. Potentially, the EU approach could evolve into a 'best practice' example for a viable regional approach to the development of a transboundary offshore wind power regime, thus effectively, a marine spatial planning approach that transcends national boundaries and which is in line with the idea of an EU internal energy market.[3]

2 Emphasizing the Multiplicity of Institutional Actors Shaping the Legal Framework

As highlighted in detail by the keynote address, the United Nations Convention on the Law of the Sea (UNCLOS)[4] provides the legal framework within which any maritime activity is performed. UNCLOS tries to strike a balance between the multiple activities that can take place simultaneously in the same maritime zone.[5] In this context, already in 2006, the United Nations General Assembly called for 'an integrated, interdisciplinary and intersectoral approach' to maritime activities as part of its annual review of the state of the oceans and the law of the sea.[6] Six years later, the United Nations General Assembly updated and substantiated its policy call by stressing public implementation challenges and 'the need to strengthen the ability of competent international organizations to contribute, at the global, regional, subregional and bilateral levels, through cooperation programmes with Governments, to the development of

3 N. Banks and S. Trevisanut, *Energy from the Sea*, Brill, 2015, p. 3.

4 United Nations Convention on the Law of the Sea (adopted 10 December 1982, entered into force 16 November 1994) 21 ILM 1261 (1982) (UNCLOS).

5 N. Banks and S. Trevisanut, *Energy from the Sea*, Brill, 2015, p. 2.

6 UNGA Res. 60/30 (8 March 2006), p. 2.

national capacity in marine science and the sustainable management of the oceans and their resources.[7]

When it comes to offshore energy generation, one has to acknowledge that there are not only public competent organizations but also a multiplicity of private institutional actors which also contribute to the shape of the relevant legal framework. Thus, not only do we have to acknowledge the important role of certain public regulators, such as the EUOAG,[8] the NSOAF[9] or the IRF[10] but also of certain industry associations, such as the International Oil and Gas Producers Association (IOGP).[11] Looking closer at the membership of the latter association, this comment to the keynote would like to highlight the vital role of private companies in ocean-related energy generation. Without private oil and gas companies, pipeline and transmission line operators, the shipping sector, the fisheries sector, and indigenous communities as well as also non-governmental organizations (NGOs), the picture would not be complete. Most coastal States neither have the scientific expertise nor the manpower to implement offshore energy generation on a larger scale. Rather, coastal States often reduce their own role (voluntarily) to making discretionary decisions on whether to authorize offshore energy projects by private companies and to exercise public competencies in prescriptive safety supervision.

The traditional 'division of labour' between coastal States and private actors (in particular oil and gas companies) has had important implications for the way in which legal rules have been developed and are being implemented in the area of offshore energy generation in the past and until today.[12] For example, potential petroleum deposits, no matter whether they fall exclusively into the functional jurisdiction of one coastal State or whether they are of a transboundary nature, will usually be developed by private licensees. For the EU, Directive 2013/30/EU[13] defines the licensee as 'the holder or joint holders of

---

7 UNGA Res. 66/231 (5 April 2012), p. 2.

8 The EUOAG was established in 2012 with Commission Decision C 18/17 (2012) OJ C 18/8. It is formed by Member States' authorities covering both offshore safety and related marine environmental protection, <http://euoag.jrc.ec.europa.eu/> accessed 1 July 2017.

9 <http://www.psa.no/nsoaf/category999.html> accessed 1 July 2017.

10 <http://www.irfoffshoresafety.com/> accessed 1 July 2017.

11 <http://www.iogp.org/Safety> accessed 1 July 2017.

12 N. Banks and S. Trevisanut, *Energy from the Sea*, Brill, 2015, p. 6.

13 Directive 2013/30/EU of the European Parliament and of the Council of 12 June 2013 on safety of offshore oil and gas operations and amending Directive 2004/35/EC, OJ L 178/66 of 28 June 2013.

a license'.[14] This entity or consortium will enter into further arrangements for unitizing[15] and for producing the resources and it will lead the relevant contractual negotiations. Above all, as a result of intensive exploration activities during the pre-authorization phase, the licensee's knowledge of the geological specifications and other properties of the deposit will often be superior as compared to the knowledge of any local public authorities.

The licensee will enter into a multitude of contractual arrangements once the decision is made that an onshore/offshore hydrocarbon discovery is commercial and warrants further development. In particular, the licensee will prepare a field development plan and submit it to the relevant governmental agency along with an application for approval to develop the field in accordance with the plan. For this reason, it has already been concluded in the relevant literature that the remaining authority of States to generally approve the umbrella of further contractual arrangements of licensees is – in reality – rather "hollow" and dependant on the performance of private actors.[16]

Consequently, non-State actors play an increasingly important role in developing the energy resources of ocean space. But non-State actors are also crucial in developing the applicable provisions, for example, when it comes to detailed performance-based safety requirements which are emanating as more safety-efficient over more prescriptive laws. A good further example for efficient industry self-regulation can be highlighted with regard to the seaborne transportation of energy resources like oil and gas. It is not only exclusively the International Maritime Organization (IMO) and its State-based membership which develops the relevant technical provisions for oil and gas tankers. In addition, marine business operators play an equally important role in developing ocean-related corporate social responsibility standards. These private standards aim at preventing marine pollution and environmental disasters and in addressing liability issues. For example, for more than 30 years now, intensive industry-led 'vetting procedures' are carried out by oil companies and by many

14 'License' means an authorisation for offshore oil and gas operations pursuant to Directive 94/22/EC of the European Parliament and of the Council of 30 May 1994 on the conditions for granting and using authorizations for the prospection, exploration and production of hydrocarbons, OJ L164/3 of 30 June 1994.

15 The term 'unitization' refers to situations where a commercial discovery extends beyond the boundary of the contract area and into the contract area that is the subject of another authorization, or an area that has yet to become the subject of another authorization. Either the relevant law or the authorization may require, or may grant to the relevant governmental agency the authority to require, that the commercial discovery be developed as a single unit.

16 N. Banks and S. Trevisanut, *Energy from the Sea*, Brill, 2015, p. 6.

other tanker charterers on a routine basis. This private standard-setting has significantly contributed to a dramatically improved global safety record of oil and gas tankers – in cooperation and coordination with public rules and related enforcement measures by flag States and port States. The complete vetting process is carried out by private entities which collect and evaluate data on tankers and who advise a system for the risk assessment of that data. In case a tanker is considered for a charter, i.e. for the seaborne carriage of hydrocarbons, the collected information will be evaluated and the outcome will be compared with a precise standard of acceptance. As a result, the tanker will either be accepted or rejected (or accepted with subjects).[17]

3 Some Thoughts on the Open Issue of Liability for (Transboundary) Offshore Accidents

This comment has appraised the complementary and sometimes leading role of private actors and private standards which generates a significant potential to improve the safety standards of offshore energy generation. Nevertheless, this appraisal shall not imply any degree of perfection. The aftermath of severe offshore disasters which occurred in the Timor Sea ('Montara', 2009)[18] and in the Gulf of Mexico ('Deepwater Horizon', 2010)[19] has clearly identified some consequences of uncoordinated deregulation and lack of administrative control. However, the impact of these disasters has also triggered administrative reforms in various regions of the world. And it is still important to assess the arguments in favour of self-regulation of the industry as well as ideas of co-regulation. Despite the global existence of the legal duty to prevent harm and the design of effective measures, including vessel design, equipment and crewing standards, as well as all applicable best practices, one significant challenge still remains in the area of offshore energy generation: This challenge relates to liability and compensation and, in particular, the creation of effective regimes to ensure that victims have access to adequate financial compensation in a timely manner.

17 T. Knowles, *Tanker Vetting*, Witherby Seamanship, 2010, p. 11.

18 See further S. Tromans, 'Pollution from Offshore Rigs and Installations', in B. Soyer and A. Tettenborn (eds.), *Offshore Contracts and Liabilities*, Informa, 2014, p. 257.

19 G. Gordon, 'The *Deepwater Horizon* disaster: the regulatory response in the United Kingdom and Europe', in R. Caddell and R. Thomas (eds), *Shipping, Law and the Marine Environment in the 21st Century: Emerging Challenges for the Law of the Sea – Legal Implications and Liabilities*, Witney: Lawtext Publishing Ltd., 2013, p. 187.

The IMO has played a leading role in developing relevant rules to deal with both prevention and civil liability in the context of vessels and especially tankers. However, there is no similar comprehensive effort to address these issues in the context of transboundary offshore oil and gas exploration and production. As a political reaction to the Montara oil spill, a joint proposal by Indonesia and Denmark was first brought to the attention of the IMO Legal Committee in 2010: This initial proposal addressed the need to develop a new Convention to address liability and compensation issues connected with transboundary pollution damages resulting from offshore exploration and exploitation activities. The matter was further intensified by the Deepwater Horizon disaster which involved, however, no transboundary pollution but highlighted, above all, the massive potential threats to the marine environment.

Between 2010 and 2016, a majority of the Members of the IMO Legal Committee (supported by the relevant industry) consistently rejected proposals to develop a new IMO Convention for this area. As a result, it is highly unlikely for the IMO to draft a new liability Convention to address the risks involved in offshore oil exploration and extraction operations which have ventured further into deeper waters and to meet the liability that arises from such activities on a multilateral level. Rather, the preferred approach is to develop a new set of guidelines to help States pursue bilateral or regional agreements in this regulatory area. At the 104th session of the IMO Legal Committee, Indonesia and Denmark jointly submitted another "guidance document" on the subject matter.[20] This recent development reinforces the necessity for regional, issue specific, and also non-State actor initiatives that also address the liability question in the context of environmental harm caused by upstream oil and gas operations. For example, since 2013, the EU has taken a much more sophisticated legal position in this regard: According to Article 4(2)(c) of Directive 2013/30/EU on safety of offshore oil and gas operations, when assessing the technical and financial capability of the applicant for a license, due account shall be taken in particular of the 'applicant's financial capabilities, including any financial security, to cover liabilities potentially deriving from the offshore oil and gas operations in question including liability for potential economic damages where such liability is provided for by national law'. Furthermore, Article 4(3) of the Directive mandates the EU Members to 'ensure that the licensing authority does not grant a license unless it is satisfied with evidence from the applicant that the applicant has made or will make adequate provision [...] to cover liabilities potentially deriving from the applicant's offshore oil and gas operations. Such provision shall be valid

20 Addressed under 'any other business', see IMO Doc. LEG/104/14/2 (21 February 2017).

and effective from the start of offshore oil and gas operations. […].' Licensees will also be required to maintain sufficient capacity to meet their financial obligations resulting from liabilities for offshore oil and gas operations.[21]

Finally, Article 7 of the Directive, in fact, governs liability for environmental damage: 'Without prejudice to the existing scope of liability relating to the prevention and remediation of environmental damage pursuant to [the EU Directive on Environmental Liability], Member States shall ensure that the licensee is financially liable for the prevention and remediation of environmental damage as defined in [the Environmental Liability Directive], caused by offshore oil and gas operations carried out by, or on behalf of, the licensee or the operator.' Thus, under the national laws of EU Members, licensees will be directly financially liable for the prevention and remediation of environmental damage caused by offshore oil and gas activities carried out by, or on behalf of, the licensee or the operator. Licensees are strictly liable under the combined legal regimes of the Offshore Safety Directive and the Environmental Liability Directive. Consequently, in transposition of Directive 2013/30/EU to the national legal orders of the EU Member States, financial liability for damages caused by offshore oil and gas operations has been strictly channelled to licensees.

4 Conclusions

The private sector plays a crucial role in the exploration and further development of offshore energy sources. The private sector actively participates and influences the law-making processes and it is also vital in the practical implementation of the relevant safety and security obligations. Additionally and on a comparative basis, the private sector often has superior technological knowledge and the financial resources to engage faster and more efficiently as compared to State actors. This situation somehow clashes with the remaining roles of coastal States and their functional sovereignty which undoubtedly includes granting authorities. No offshore-related energy generation activity can take place without the explicit authorization(s) of the coastal State(s) under whose jurisdiction the project will be realized. This situation has been correctly identified as a 'diffuse authority' in the energy sector: The authority, and the

21 It can be expected that the majority of EU Member States will implement Article 4(3) of the Directive by requiring a proof of insurance from licensees (with minimum standards for coverage, as some national legal orders within the EU do already prescribe).

rights and obligations related thereto, are shared between different actors (States and private actors) often also at different levels of governance (local, national, regional, international, etc.).[22]

In sum, this situation generates both advantages of greater flexibility but also risks, such as an inappropriate priority-setting between economic interests and ecologic requirements. In any case, all stakeholders of the civil society which are interested in sustainable ocean energy generation will be in further need to substantiate our common understanding of the concept of shared responsibilities. A lot of contributions to the Third Hamburg International Environmental Law Conference of 2016 will definitely shed some further light on this important global policy challenge.

Bibliography

Banks, N. and Trevisanut, S., *Energy from the Sea*, Brill, 2015.

Gordon, G., 'The *Deepwater Horizon* disaster: the regulatory response in the United Kingdom and Europe', in R. Caddell and R. Thomas (eds), *Shipping, Law and the Marine Environment in the 21st Century: Emerging Challenges for the Law of the Sea – Legal Implications and Liabilities*, Witney: Lawtext Publishing Ltd., 2013.

Knowles, T., *Tanker Vetting*, Witherby Seamanship, 2010.

Tromans, S., 'Pollution from Offshore Rigs and Installations', in B. Soyer and A. Tettenborn (eds.), *Offshore Contracts and Liabilities*, Informa, 2014.

22 For further details see N. Banks and S. Trevisanut, *Energy from the Sea*, Brill, 2015, p. 11.

CHAPTER 4

International Environmental Law, Sustainable Generation of Energy from the Ocean and Small Island Developing States in the Pacific

David Kenneth Leary

1 Introduction

Pacific Island countries and territories (PICTs) are heavily dependent upon imported fossil fuel to meet their energy needs.[1] The combined effect of low GDP, high energy prices, small population density and remoteness means that PICTs are extremely vulnerable to external energy crises.[2] The importance of energy security for PICTs and the role renewable energy can play in providing such security in a sustainable manner has been recognized in a range of soft law instruments and other international initiatives over the past few decades.

While there has been developed a range of different technologies that harness energy from the oceans, most research has focused on three key areas: (1) hydrokinetic energy, where the energy of ocean currents and tides are captured by devices which are installed under the surface of the water; (2) wave energy, where the energy of the surface wind waves is used to produce electricity by a variety of devices installed on the surface of the Sea; and (3) Ocean thermal Energy conversion or OTEC, which utilizes the temperature differential between cold water from the deep ocean and warm surface water.[3] This paper considers the potential role that these technologies could play in sustainably meeting the energy needs of PICTs.

1 A. Woodruff, 'An Economic Assessment of Renewable Energy Options for Rural Electrification in Pacific Island Countries', *SOPAC Report 397*, 2007.

2 M. Jafar, 'Renewable Energy in the South Pacific-Options and Constraints', *Renewable Energy*, vol. 19, 2000, p. 305.

3 H. Soerensen and A. Weinstein, 'Ocean Energy: Position Paper for IPCC Proceedings of IPCC Scoping Meeting on Renewable Energy Sources', Lubeck 2008. For a detailed review of the various ocean energy technologies see A. Lewis et al., '*Ocean Energy*', in O. Edenhofer et al., *IPCC Special Report on Renewable Energy Sources and Climate Change Mitigation*, Cambridge University Press, 2011.

 | DOI 10.1163/9789004360273_005

While traditionally debates on sustainable generation of energy from the oceans has focused on offshore wind energy in Europe, parts of Asia and North America, and ocean, tidal and wave energy in Europe and North America, this paper will highlight that there has also been considerable interest in the potential for sustainable energy generation from the oceans in the Pacific region. However, despite this interest there are few examples where the regions potential has been realized and this paper considers why. The paper begins by firstly situating developments in relation to marine renewable energy in the Pacific in the context of developments in international law and policy relating to renewable energy and small island developing States. This first part of the paper examines developments in 'soft law' over the past few decades. This is then followed by a brief examination of the relevant provisions of the 1982 *United Nations Convention on the Law of the Sea* (UNCLOS).[4] The paper then goes on to examine work on marine renewable energy that have been undertaken in the region in the past few decades. The final part of the paper then considers key barriers to the realization of the regions potential.

2 Background and Context – Renewable Energy since Stockholm

By the late 1960s into the early 1970s, a period that arguably represents the emergence of modern international environmental law, serious concerns were being expressed about growing evidence of environmental degradation, and it was realized that this degradation was closely linked to unsustainable levels of economic development and population growth. By the time of the lead up to the United Nations Conference on the Human Environment in Stockholm in 1972, an understanding was emerging that development and environmental issues were clearly inter-dependent. The meeting of the United Nations Conference on the Human Environment held in Stockholm in 1972 was a pivotal moment in the development of modern international environmental law with the *Stockholm Declaration*[5] laying the foundations for many of the key principles of international environmental law that we take for granted today.

While the *Stockholm Declaration*[6] did not refer specifically either to development and small island developing States or renewable energy, many of the

4 United Nations Convention on the Law of the Sea (adopted 10 December 1982, entered into force 16 November 1994) 21 ILM 1261 (1982) (UNCLOS).

5 Declaration of the United Nations Conference on the Human Environment (adopted 16 June 1972) UN Doc. A/CONF. 48/14/ Rev. 1 (1972) (Stockholm Declaration).

6 Ibid.

key principles of international environmental law and policy that have guided approaches to both are clearly present in this pivotal soft law document. For example, in Principle 1 of the *Stockholm Declaration* we see a clear articulation of what we now know as the Principle of intergenerational equity. Perhaps more importantly, in Principle 20 of the *Stockholm Declaration* we see explicit reference to the importance of technology transfer from developed to developing countries to assist in dealing with environmental problems.

By the time of the Earth Summit in Rio de Janeiro in 1992 the link between development and addressing global environmental challenges was clearly a mainstream concept. Thus consistent with the concepts articulated in the *Stockholm Declaration*, the *Rio Declaration*[7] in Principle 3 clearly articulates the principal of intergenerational equity and likewise scientific and technological transfer is again clearly recognized in Principle 9. The big change though between the Stockholm and the Rio meeting in 1992 was that these and many other principles were now firmly established and recognized as part of what we now know as the concept of sustainable development.

More importantly, it is at the Earth Summit in Rio de Janeiro in 1992 that we see explicit reference to the development and environmental issues in the context of small island States in *Agenda 21*.[8] In addition, and perhaps more importantly for present purposes, *Agenda 21* clearly outlined a vision for the potential role of renewable energy in meeting the energy needs of developing countries, including small island developing States. *Agenda 21* articulated a number of key themes with respect to renewable energy that have been repeated in numerous international instruments since. Thus paragraph 7.51 of *Agenda 21*[9] called for a comprehensive approach to human settlement and development which was to include the promotion of sustainable energy development in all countries. Specifically, it called upon developing countries to

> formulate national action programs to promote integrated development of energy saving and renewable energy technologies, particularly for the use of solar, hydro, wind and biomass sources [and] promote wide dissemination and commercialisation of renewable energy technologies through suitable measures, inter alia, fiscal and technology transfer mechanisms [and] carry out information and training programs directed

7 *Rio Declaration on Environment and Development* (adopted 13 June 1992) UN Doc. A/CONF.151/26/Rev. 1 96, 31 ILM 874 (1992) (Rio Declaration).

8 Agenda 21, United Nations Conference on Environment and Development, Rio de Janeiro, Annex, Resolution 1, UN Doc. A/CONF.151/26/Rev. 1, vol. 1 (1993) (Agenda 21).

9 Ibid.

> at manufacturers and users in order to promote energy saving techniques and energy efficient appliances.[10]

In that context *Agenda 21* then called upon international organisations and bilateral donors to

> support developing countries in implementing national energy programs in order to achieve widespread use of energy-saving and renewable energy technologies, particularly the use of solar, wind, biomass and hydro sources [and] provide access to research and development results to increase energy use efficiency levels in human settlements.[11]

Similarly, paragraphs 9.9 to 9.12 of *Agenda 21* devoted considerable attention to both renewable energy and the development of measures to encourage greater energy efficiency in both developed and developing countries alike. This involved a number of key elements including the development of national action plans, the important role of capacity building in the context of renewable energy, technology transfer from developed to developing countries, and most importantly finance mechanisms to facilitate the growth of renewable energy in developing countries.

Two years later the United Nations Global Conference on the sustainable development of small island developing States was held in Barbados. This meeting adopted the *Barbados Declaration*[12] and the *Barbados Plan of Action for Small Island Developing States* (BPOA).[13] The BPOA acknowledged explicitly that the development and use of renewable sources of energy and the dissemination of sound and efficient energy technologies have a central role to play in mitigating the adverse impact of climate change.[14] Similarly it acknowledged that small island developing States were heavily dependent on imports of petroleum products that are used in electricity generation and transport and that this will continue in the short to medium term.[15] The BPOA also

---

10 Ibid., para 7.5.1(a).

11 Ibid.

12 Report of the Global Conference on the Sustainable Development of Small Island Developing States, Bridgetown, Barbados, April 25 - May 6 1994, UN Doc. A/CONF.167/9 (1994) (Declaration of Barbados), Annex 1.

13 Ibid., Annex 2: Barbados Programme of Action for the Sustainable Development of Small Island Developing States.

14 Ibid., para. 19.

15 Ibid., para. 35 and 36.

acknowledged the different scale of renewable energy resources amongst individual small island developing States. Thus in setting out the basis for action the BPOA noted:

> All have substantial resources, which have still not been developed to their full potential. Wind potential is highly variable with location, both within and between countries. Hydroelectric power is a possibility only for some islands. Biomass endowment is common but unequal. Studies of the potential to geothermal, ocean thermal energy conversion and wave energy are continuing.[16]

Importantly the BPOA also recognized the limitations on the full utilisation of renewable energy by small island developing States. Thus it observed:

> Several constraints to the large-scale commercial use of renewable energy resources remain, including technology development, investment costs, available indigenous skills and management capabilities. Small-scale application for rural electrification has been sporadic. The use of renewable energy resources as substantial commercial fuels by small island developing states is dependent on the development and commercial production of appropriate technologies.[17]

Accordingly, the BPOA then set out a plan of action at the national regional and global levels for island developing States to move from their heavy dependence on imported energy sources towards a greater utilisation of their indigenous renewable energy resources. At the national level the BPOA committed small island developing States to implementing appropriate public education and awareness programs and consumer incentives to promote energy conservation.[18] Small island developing States also committed to promote the efficient use of energy, the development of environmentally sound sources of energy, energy efficient technologies and, in particular, the development of economic instruments and incentive structures to increase the economic possibility of the use of renewable energy.[19] Small island developing States similarly committed to the establishment and strengthening, where appropriate, of research capabilities in the development and promotion of new and

16 Ibid., para. 37.
17 Ibid., para. 38.
18 Ibid., Chapt VII, A (i).
19 Ibid., Chapt VII, A (ii).

renewable sources of energy, including wind, solar, geothermal, hydro-electric, wave and biomass energy, and OTEC.[20] In addition the BPOA recognized the need for action at the regional level.

Likewise the BPOA recognized the need to establish and strengthen research and policy capabilities in the development of new and renewable sources of energy including wind, solar, geothermal, hydroelectric, wave and biomass.[21] At the regional level countries were called upon to assist, where appropriate, in the formulation of energy policies, standards and guidelines for the energy sector that are applicable to small island developing States, and enhance national capacity to effectively plan, manage and monitor their energy sectors.[22] Countries also undertook on a regional basis to gather and disseminate information, and promote regional cooperation regarding technical exchanges among small island developing States' energy-sector issues, including new and renewable sources of energy.[23]

At the global level the BPOA called on States to support the research, development and utilisation of renewable sources of energy and related technologies and improve the efficiency of existing non-renewable energy technologies.[24] States also committed to develop effective mechanisms for technology transfer and dissemination of information on experience in the use of new and renewable sources of energy.[25] Finally, the BPOA also called upon international institutions and agencies, including public international finance institutions, to incorporate energy efficiency and conservation principles into energy sector related projects, training and technical assistance, and where appropriate, to provide concessionary financing facilities for energy sector reforms.[26]

Thus by the Barbados meeting in 1994, that is some 23 years ago, we could clearly discern a number of key elements in the international community's recognition of the role of renewable energy in the sustainable development of small island developing States including:

- a recognition of the economic as well as the environmental imperative of small island developing States moving away from fossil fuels to sustainable energy;

---

20 Ibid., Chapt VII, A (iii).
21 Ibid., Chapt VII, B (i).
22 Ibid., Chapt VII B (ii).
23 Ibid., Chapt VII B (iii).
24 Ibid., Chapt VII C (i).
25 Ibid., Chapt VII C (iii).
26 Ibid., Chapt VII C (iv).

- acknowledgement that the available renewable resources vary between small island developing States and therefore that future policy responses needed to be tailored to the needs, resources and capacity of individual States;
- that there was a need for each small island developing States to develop policies and strategies relating to renewable energy unique to their individual needs;
- the need for technology transfer and capacity development; and
- the need for financial assistance from developed countries to assist small island developing States and in the way from fossil fuels.

Eleven years later, in 2005, the international meeting to review the implementation of the programme of action for the sustainable development of small island developing States largely echoed these key elements by restating them in the *Mauritius Declaration*[27] and the *Mauritius strategy for the further implementation of the programme of action or the sustainable development of small island developing States* (*Mauritius strategy*).[28] One significant addition though was the greater prominence given to the potential role of public and all public/private partnerships, and market oriented approaches. This change reflected in part the increased emphasis on market based mechanisms in many international forums and instruments dealing with environmental issues over the previous decade.

More recently, in 2010, the 5-year review of the *Mauritius Strategy* significantly acknowledged the great progress that had been made by many small island developing States in developing national energy plans, policies or action plans as well as the adoption of renewable energy targets. In the specific context of the Pacific that praise is warranted. Most of the small island developing States and territories in the Pacific region have already adopted very ambitious renewable energy targets towards the year 2020 or have otherwise adopted ambitious targets not necessarily linked to the year 2020.[29] Likewise

---

27 Report of the International Meeting to Review the Implementation of the Programme of Action for the Sustainable Development of Small Island Developing States, Port Louis, Mauritius, 10–14 January 2005, UN Doc A/CONF.207/11 (2005) (Mauritius Declaration), Annex 1.

28 Ibid., Annex II: Mauritius Strategy for the Further Implementation of the Programme of Action for the Sustainable Development of Small Island Developing States.

29 For specific details of these targets and other policies see International Renewable Energy Agency, Policy Challenges for Renewable Energy Deployment in Pacific Island Countries and Territories, 2012, <https://www.irena.org/DocumentDownloads/Publications/Policy_Challenges_for_Renewable_Energy_Deployment_PICTs.pdf.> accessed 1 July 2017.

as the international renewable energy agency has highlighted in its recent report *Pacific Lighthouses Roadmapping for Islands*[30] there is now an extensive body of policy documents including roadmaps, strategies, and action plans, directed towards achieving the significant aspirational goals that have been set by Pacific Island countries and territories in the Pacific region.

Most recently, in 2016, in the context of Goal 7 of the *2030 Agenda for Sustainable Development*[31] (ensuring access to affordable, reliable, sustainable and modern energy for all), on a non-legally binding basis States have agreed by 2030 to

> expand infrastructure and upgrade technology for supplying modern and sustainable energy services for all in developing countries, in particular least developed countries, small island developing States and landlocked developing countries, in accordance with their respective programmes of support.[32]

3 The Law of the Sea and Existing Institutions

Before turning to examine developments relating to ocean energy in the Pacific it is first useful to turn briefly to examine the relevance of provisions of UNCLOS. In this context the provisions of UNCLOS dealing with the rights of the coastal State with respect to its internal waters, territorial sea, and the exclusive economic zone clearly articulate the rights and obligations of both the coastal State and any users of those waters. It is clearly settled law that the coastal State has full sovereignty over its internal waters and its territorial sea, and therefore has jurisdiction to regulate all aspects of the installation, operation, and decommissioning of various renewable energy installations in such ocean space.[33]

It is also clearly settled law that regard also has to be had to the right of innocent passage of all ships through internal waters under Article 8(2) of UNCLOS and the territorial sea under Article 17 of UNCLOS. In the territorial sea, for

30 International Renewable Energy Agency, Pacific Lighthouses Renewable Energy Roadmapping for Islands, (2013), <https://www.irena.org/DocumentDownloads/Publications/Pacific-Lighthouse-Roadmapping.pdf> accessed 1 July 2017.

31 UNGA Res. 70/1 'Transforming Our World: The 2030 Agenda for Sustainable Development' (adopted 25 September 2015).

32 Ibid., Goal 7a.

33 See in particular the provisions of Article 2(1) of UNCLOS.

example, it is a reasonable question to ask whether renewable energy installations are permissible if they potentially interfere with rights of innocent passage? As has been argued elsewhere,[34] the law is quite clear in this regard. Consideration of the right of innocent passage needs to be balanced against the explicit recognition of the rights of the State to regulate aspects of its passage contained in Articles 21 and 22 UNCLOS.[35]

Similarly, the limited experience we have with renewable energy generation from the oceans suggests international institutions are well equipped to sustainably manage this emerging technology. For example, the *International Maritime Organisation* (IMO) worked closely with the United Kingdom in implementing changes to regulation of navigation of vessels in the waters off Land's End, between the United Kingdom mainland and the Isles of Scilly in South West England to accommodate the installation of the wave energy project known as the Wave Hub. There the United Kingdom in conjunction with the IMO developed amendments to the existing traffic separation scheme for vessels in these waters to ensure the safety of navigation and protection of the marine environment.[36]

It is not possible to consider these issues in detail in this short paper, but this example does illustrate the point that to date existing international institutions are well equipped to respond to the challenges of this emerging technology. Having said that, it is noted however, that some scholars have argued that there is a need for greater international regulation of renewable energy generation in the oceans in areas beyond national jurisdiction. Concerns in particular have been expressed about the lack of regulation of OTEC generation technologies and technologies utilising osmotic energy, which relates to the pressure differential between saltwater and freshwater. But both OTEC and osmotic energy are still highly experimental and we are unlikely to see any significant utilisation of this technology in the ocean space beyond national jurisdiction in the foreseeable future. The main forms of renewable energy generation from the ocean which have been installed, either as pilot plants or as full-scale commercial generating facilities such as tidal energy plants, wave energy installations or offshore wind farms have been installed either within the internal waters, territorial sea, or exclusive economic zones of coastal States. In these jurisdictional zones the adequacy of existing international law and institutions does not appear to be in question. Instead, as is outlined

---

34 See D. Leary and M. Esteban, 'Recent Developments in Offshore Renewable Energy in the Asia-Pacific Region', *Ocean Development & International Law*, vol. 42, 2011, pp. 94–119.

35 Ibid.

36 This example is considered in more detail in ibid.

below, it does appear that it is the capacity of existing domestic laws and a range of other factors that present barriers to future marine renewable energy projects.

4 The Challenges Faced in Realizing the Potential of Renewable Energy Generation from the Oceans

As the above discussion highlights, there have been an extensive range of soft law instruments that recognize the important role renewable energy can play in providing energy security in the future for small island countries and territories. The following discussion considers the extent to which the aspirational goals of the soft law instruments discussed above has been realized in the waters over which PICTS have jurisdiction as recognized by UNCLOS, with a particular focus on sustainable energy generation from the oceans.

The experience in the Pacific region highlights the significant obstacles faced by small island developing States in taking advantage of the immense potential for renewable energy generation from the ocean. Attempts to exploit the renewable energy potential of the oceans have been under way in the Pacific region for several decades. As early as 1981 a trial of OTEC technology was undertaken in Nauru. From 1981 to 1982 the Tokyo Electric Power Company and Toshiba undertook technical trials of a mini-OTEC facility in Nauru and this was in fact the world's first OTEC plant to feed electricity into an operational electricity grid.[37] As recently as 2008 Japanese company Xenesys Inc[38] started a joint venture to conduct commercial and technical feasibility studies for OTEC in French Polynesia, New Caledonia and Vanuatu.

More significantly, there has been a number of studies of the potential of tidal and wave energy in the Pacific region with the earliest study for tidal energy for example dating back to a 1976 feasibility study undertaken in the Cook Islands.[39] Likewise the potential of wave energy in Pacific Island countries has been the subject of several detailed studies including by regional intergovernmental organisations such as the Secretariat of the Pacific Community

37 H. Wade et al., *Pacific Regional Energy Assessment: An Assessment of the Key Energy Issues, Barriers to the Development of Renewable Energy to Mitigate Climate Change, and Capacity Development Needs for Removing the Barriers*, 2004.

38 <http://xenesys.com/english/> accessed 1 July 2017.

39 H. Wade et al., *Pacific Regional Energy Assessment: An Assessment of the Key Energy Issues, Barriers to the Development of Renewable Energy to Mitigate Climate Change, and Capacity Development Needs for Removing the Barriers*, 2004.

through its predecessor the *South Pacific Geosciences Commission* (SOPAC), the *South Pacific Regional Environment Program* (SPREP) and projects funded by international aid agencies such as the *Norwegian Agency for International Development*.[40] The latter study, for example, undertaken from 1987 to 1995, involved a wave energy resource assessment undertaken in the Cook Islands, Fiji, Tonga, Tuvalu, Vanuatu and Samoa.

While there have been feasibility studies of energy generation in the oceans in the Pacific, there is still a lack of data on the nature and scale of the resources in individual Pacific Island countries. Indeed, much of information that does exist is contained in governmental and intergovernmental reports that are difficult to access, and little useful data is available in the mainstream scientific literature. Having said, that a number of such studies are worth noting. Firstly, a report published in 2001 by SOPAC suggests the Cook Islands, Fiji, Guam, Kiribati, New Caledonia, Samoa, the Solomon Islands and Vanuatu have the potential for OTEC plants.[41] Likewise a study of renewable energy in Pacific Island countries undertaken by SOPAC in 1998 vaguely noted potential for development of OTEC as 'good' in the Cooks Islands, Fiji, Kiribati, the Marshall Islands and Nauru.[42] The same report also noted that Fiji and Tonga possessed 'good' wave resources while also noting 11 other Pacific Island countries, including the Federated States of Micronesia, Kiribati, the Marshall Islands, Nauru, Niue, Palau, PNG, Samoa, Solomon Islands, Tuvalu and Vanuatu, were all assessed as having definite potential, but the extent of wave energy resources was noted as unknown. More recently in 2004 and 2005, the *Pacific Islands Renewable Energy Project* (a joint project of SPREP, GEF, UNDP and Pacific Island countries) published a series of reports identifying opportunities for renewable energy in Pacific Island countries with some observations on the potential of renewable energy generation from the oceans in Pacific Island countries. A subsequent SOPAC study in 2009 further highlighted the potential of ocean based renewable energy without offering any clear insight on the region's best opportunities for taking advantage of the perceived resource.

As good as these various assessments and studies have been they are in many respects simplistic and the use of descriptive terms such as 'good' is very vague. It was not clear on the face of the studies what was the scientific

40 See Scottish Enterprise, *Marine Renewable (Wave and Tidal) Opportunity Review*, 2005.

41 R. Mario, 'Ocean Thermal Energy Conversion and the Pacific Islands', *SOPAC Miscellaneous Report*, vol. 417, 2001.

42 P. Fairburn, 'A Regional View Towards Sustainable Renewable Energy Development in the Pacific', *SOPAC Miscellaneous Report*, vol. 311, 1998.

basis of qualitative conclusions such as these. Thus while there has been much speculation as to the potential of renewable energy generated from the Pacific Ocean region even a review of reports coming from the numerous international initiatives in the region suggests there is a clear need for more detailed scientific research to clearly identify specific locations in specific countries where commercial or even pilot scale projects could be developed. These studies would then need to be followed by further studies of the economic and technical feasibility of the exploitation of such resources. Such research will be expensive and it is clear that Pacific Island countries have neither the financial, nor the technical and scientific expertise to undertake such research without assistance from developed countries.

Even if there was sufficient scientific data available on the scale of resources and the feasibility of their exploitation, a range of previous studies have clearly identified that there are significant barriers that Pacific Island countries need to overcome before they are able to exploit these potential resources. These include: a lack of awareness and knowledge of existing energy options;[43] limited expertise in design, installation, operation and maintenance of renewable energy systems;[44] poor policy planning;[45] legal and regulatory barriers including inadequate legal frameworks, including onerous grid connection requirements for small power producers set by utility entities;[46] high capital costs;[47] existing foreign debt burdens which raise problems in obtaining finance for projects;[48] a lack of access to capital to develop such projects with renewable energy projects competing with other pressing priorities;[49] lack of subsidies for

43 D. Weisser, 'On the Economics of Electricity Consumption in Small Island Developing States: A Role for Renewable Energy Technologies', *Energy Policy*, vol. 32, 2004, p. 127.

44 P. Johnston, 'Pacific Regional Energy Assessment: An assessment of the key energy issues, barriers to the development of renewable energy to mitigate climate change, and capacity development needs for removing the barriers'-*Fiji Natonal Report-SPREP*, 2005.

45 A. Woodruff, 'An Economic Assessment of Renewable Policy Options for Rural Electrification in Pacific Island Countries', *SPOAC Technical Report*, 2007, p. 397.

46 T. Urmee et al., 'Issues related to Rural Electrification using Renewable Energy in Developing Countries of Asia and the Pacific', *Renewable Energy*, vol. 34, 2009, p. 354.

47 Ibid.

48 D. Weisser, 'On the Economics of Electricity Consumption in Small Island Developing States: A Role for Renewable Energy Technologies', *Energy Policy*, vol. 32, 2004, p. 127.

49 X. Yu and R. Tapling, 'Policy Perspectives: Environmental Management and Renewable Energy in the Pacific Island', *Journal of Environmental Management*, vol. 51, 1997, p. 107.

renewable energy;[50] inadequate institutional and human resource capacities;[51] markets that are too small to attract foreign investment in renewable energy projects;[52] lack of economies of scale[53] and size of electricity generation utility companies;[54] limited access to technology appropriate to the needs of these countries;[55] foreign investor perceptions;[56] and perhaps more critically given the dominance of international aid funding for such projects, a lack of coordination in aid projects and programs.[57]

Perhaps one of the most significant of these barriers is access to funding for any future utility scale projects. Access to both public and private sector funding will be critical for any future renewable energy projects in the Pacific. Experience of projects such as wave and tidal energy projects in developed countries shows that significant government and private sector investment is required to make such projects a reality. However, small island States in the Pacific region do not have access to the required financial resources. Any future development of renewable energy generation from the oceans in these countries will be heavily dependent upon overseas development assistance from developed countries and multilateral lending agencies who often have their own priorities. Although we have seen a small amount of foreign aid assisting with some of the early studies of OTEC, wave and tidal energy feasibility in the Pacific region, for the foreseeable future we are likely to see aid donors focusing on more mature technology such as solar energy (principally photovoltaics), biofuels and small-scale on shore wind energy. The elephant in the room, so to speak, is why would aid donors invest in the as yet largely experimental technology of marine renewables when trusted and proven technology such as solar PV cells and small scale wind energy is readily available and already has a track record of success in the Pacific region. In that respect it is worth noting

50 T. Urmee et al., 'Issues related to Rural Electrification using Renewable Energy in Developing Countries of Asia and the Pacific', *Renewable Energy*, vol. 34, 2009, p. 354.

51 D. Weisser, 'On the economics of electricity consumption in small island developing states: a role for renewable energy technologies', *Energy Policy*, vol. 32, 2004, p. 127.

52 A. Woodruff, 'An Economic Assessment of Renewable Policy Options for Rural Electrification in Pacific Island Countries', *SPOAC Technical Report*, 2007, p. 397.

53 Ibid.

54 T. Roper, 'Small Island States – Setting an Example on Green Energy Use', *Review of European Community and International Environmental Law*, vol. 14, 2005, p. 108.

55 A. Woodruff, 'An Economic Assessment of Renewable Policy Options for Rural Electrification in Pacific Island Countries', *SPOAC Technical Report*, 2007, p. 397.

56 Ibid.

57 Forum for Energy and Development, *Proceedings from the Global Conference on Renewable Energy Islands*, Aero, Denmark, 1999.

that no recent project in the Pacific region has dedicated significant amounts of funding for renewable energy generation from the oceans.

A related issue that has been identified in developed countries is the need for feed-in tariffs as a mechanism to provide additional financial incentive for investment in renewable energy generation from the ocean.[58] It is questionable if any Pacific Island country has either the financial assets or the existing legislative mechanisms for feed-in tariffs to encourage investment in ocean energy in their respective jurisdictions. The potential of import duties, taxes, licenses and concession fees and other government charges to act as barriers to deployment of renewable energy technologies also stands out as another significant barrier to the implementation of renewable energy generation from the oceans in the Pacific.

With aid funding for renewable energy projects being set by donor priorities in the Pacific, there is also little incentive for private sector involvement in renewable energy projects. In the region, there has been a complete lack of involvement of the private sector in renewable energy service delivery. But there are a number of mechanisms that have already been successful in other jurisdictions that the international community might want to look at closely in seeking to provide appropriate incentives for business to invest, or alternatively to assist households or communities in investment in renewable energy technology. Many of the incentives and financial mechanisms that have been adopted in developed countries to promote renewable energy cannot easily be translated to developing countries because of the very different economic and development conditions. However, the experience of other developing countries in providing such mechanisms does show this can be done. Jurisdictions such as East Timor, Indonesia, the Philippines and Thailand (all developing countries) have all adopted a range of different measures to help provide incentives for investment by business and by households and communities in renewable energy. In East Timor's case, for example, there have been income tax, sales tax and customs duty exemptions for investments in renewable energy projects. Similarly, the Philippines and Thailand have introduced a feed-in tariff for different forms of renewable energy and in Indonesia similar proposals are currently under development and will be implemented for different forms of renewable energy technology.[59]

---

58 D. Leary and M. Esteban, 'Recent Developments in Offshore Renewable Energy in the Asia-Pacific Region', *Ocean Development & International Law*, vol. 42, 2011, p. 94.

59 Details of these measures are available from the joint International Energy Agency/International Renewable Energy Agency Joint Policies and measures database <http://www.iea.org/policiesandmeasures/renewableenergy/> accessed 1 July 2017.

International experience, especially in more developed countries such as Germany and other jurisdictions within the European Union, suggests that feed-in tariffs are one of the most useful incentives to further renewable energy development. Feed-in tariffs are in essence, guaranteed prices for electricity supply to cover the cost of generation plus a reasonable profit to induce developers to invest in renewable energy. They provide a premium for any excess electricity that may be fed back into the electricity network from renewable energy sources.[60]

Another incentive mechanism worth noting are a series of subsidies that assist individual households, communities and businesses to adopt renewable energy technologies. For example, in Palau the National Development Bank of Palau, in conjunction with the Palau energy office, has developed a number of different programs which are currently being implemented that assist households in moving towards greater adoption of renewable energy. These include a renewable energy subsidy loan program which finances the purchase of both on grid and off grid photovoltaic cells as well as solar hot water heaters.[61] Other programs also assist with energy efficiency measures.[62] While Palau is in a somewhat different position to many parts of the Pacific because of its free Association with the United States of America, this does provide an interesting example and perhaps merits a closer study of how financing mechanisms might work in other Pacific Island countries and territories. A range of measures like these will be important in assisting island States in the adoption of renewable energy technology specific to their individual needs.

Finally, a significant legal barrier to the development of renewable energy generation from the ocean which is a problem faced by developed and developing countries alike is the lack of clear regulatory frameworks for such projects. Clearly there is a need for law reforms in both developed and developing countries, especially with respect to existing legislative frameworks for

---

60 National Energy Regulator of South Africa, 'South Africa Renewable Energy Feed-In Tarriff (REFIT): Regulatory Guidelines', 26 march 2009, <http://www.sessa.org.za/component/docman/doc_download/53-nersa-media-release-renewable-energy-feed-in-tariff-refit> accessed 1 July 2017.

61 For a detailed examination of Palau's measures see K. Udui et al., 'Case study 1 – Capacity Building for Financing Institutions – Establishing Energy Loan Program', 2014, <http://www.irena.org/documentdownloads/events/2015/palau%20financing%20training/Case%20study_Energy%20loan%20for%20RE%20Palau%20model.pdf> accessed 1 July 2017.

62 For further details on programs in Palau see Energy Transition Initiative, 'Energy Snapshot-Palau' <http://www.nrel.gov/docs/fy15osti/64291.pdf> accessed 1 July 2017.

assessing the potential environmental impact of renewable energy generation from the oceans. In most jurisdictions to the extent that there is clear regulation the existing regulatory frameworks are not fit for purpose.

5 Conclusion

In conclusion, while the oceans may potentially provide an important sustainable source of renewable energy in the Pacific, it is unlikely that this potential will be realized on any significant scale in the near future. There are a wide range of barriers to adoption of this technology in PICTs, not the least of which is the lack of proper understanding of the scale and potential of renewable energy from the oceans in the region.

In terms of international law and policy it is clear that over the past two decades renewable energy has been recognized as a key pathway to energy security for small island developing States and for PICTs in particular. More significantly UNCLOS provides a clear framework recognizing the rights and obligations of PICTs in waters where marine renewable energy projects could potentially be installed. International law of the sea and existing international institutions appear to be more than adequate to address the challenges that may arise from sustainable energy generation from the oceans. In many respects the issue is not the absence of international environmental law, but rather the absence of proper and meaningful implementation of the soft law obligations that have solidified over the past few decades. The challenge for nation States is to move beyond soft law statements, feasibility studies and data collection and instead move to implementing concrete measures to assist PICTs to realize the potential that their renewable energy resources offer for energy security and sustainable development more broadly.

Bibliography

Fairburn, P., 'A Regional View Towards Sustainable Renewable Energy Development in the Pacific', *SOPAC Miscellaneous Report*, vol. 311, 1998.

Jafar, M., 'Renewable Energy in the South Pacific-options and constraints', *Renewable Energy*, vol. 19, 2000, pp. 305–309.

Johnston, P. 'Pacific Regional Energy Assessment: An assessment of the key energy issues, barriers to the development of renewable energy to mitigate climate change, and capacity development needs for removing the barriers'-*Fiji Natonal Report-SPREP*, 2005.

Leary, D. and Esteban, M., 'Recent Developments in Offshore Renewable Energy in the Asia-Pacific Region', *Ocean Development & International Law*, vol. 42, 2011, pp. 94–119.

Lewis, A. et al., '*Ocean Energy*', in O. Edenhofer et al., *IPCC Special Report on Renewable Energy Sources and Climate Change Mitigation*, Cambridge University Press, 2011.

Mario, R., 'Ocean Thermal Energy Conversion and the Pacific Islands', *SOPAC Miscellaneous Report*, vol. 417, 2001.

Roper, T., 'Small Island States – Setting an Example on Green Energy Use', *Review of European Community and International Environmental Law*, vol. 14, 2005, pp. 108–116.

Scottish Enterprise, *Marine Renewable (Wave and Tidal) Opportunity Review*, 2005.

Soerensen, H. and Weinstein, A., 'Ocean Energy: Position paper for IPCC Proceedings of IPCC Scoping Meeting on Renewable Energy Sources', Lubeck, Germany 2008.

Urmee, T. et al., 'Issues related to rural electrification using renewable energy in developing countries of Asia and the Pacific', *Renewable Energy*, vol. 34, 2009, pp. 354–357.

Wade, H. et al., *Pacific Regional Energy Assessment: An assessment of the key energy issues, barriers to the development of renewable energy to mitigate climate change, and capacity development needs for removing the barriers*, Secreteriat of the Pacific Regional Environment Programme (SPREP), Apia, Samoa, 2004.

Weisser, D., 'On the economics of electricity consumption in small island developing states: a role for renewable energy technologies', *Energy Policy*, vol. 32, 2004, pp. 127–140.

Woodruff, A., 'An economic assessment of renewable energy options for rural electrification in Pacific Island countries', *SOPAC Report 397*, 2007.

Yu, X. and Tapling, R., 'Policy perspectives: Environmental Management and Renewable Energy in the Pacific Island', *Journal of Environmental Management*, vol. 51, 1997, pp. 107–122.

CHAPTER 5

Realization of Sustainable Management/ Development under the Law of the Sea Convention?

Rüdiger Wolfrum

1 Introduction

Excellencies, colleagues, ladies and gentlemen I intend to touch upon a topic which – although couched in general, perhaps even academic, terms – has a very concrete background: namely the interrelationship between the regime based upon the Biodiversity Convention and the regime in the UN Convention on the Law of the Sea (UNCLOS)[1] concerning marine genetic resources. I am aware of the ongoing negotiations on marine genetic resources in areas beyond national jurisdiction. As the title of my presentation indicates, I shall only deal with this issue from the perspective of sustainability and whether this aspect is covered by or can be brought under the umbrella of UNCLOS.

Why has the issue of marine genetic resources come up with regard to UNCLOS and why is there this movement – endorsed by the UN General Assembly – to look for a new legal regime under UNCLOS concerning the management of marine genetic resources? To put it in a nutshell, new scientific findings, insights and developments have been generated concerning genetic resources[2] which are reflected in the Biodiversity Convention as well as in its Protocol on Access to Genetic Resources and the Fair and Equitable Sharing of Benefits Arising from their Utilization (Nagoya Protocol) and other international instruments. Marine genetic resources are not explicitly referred to

* The views expressed in this presentation are strictly made in my personal capacity; they may not be attributed to ITLOS or be taken so as to reflect a policy of the Tribunal.

1 United Nations Convention on the Law of the Sea (adopted 10 December 1982, entered into force 16 November 1994) 21 ILM 1261 (1982) (UNCLOS).

2 Article 2 of the Convention on Biological diversity provides definitions of the terms of genetic resources and genetic material and serves as a starting point for a better understanding of marine genetic resources. According to this article "genetic material" means 'any material of plant, animal, microbial or other origin containing functional units of heredity'. "Genetic resources" means 'genetic material of actual or potential use'. According to these definitions genetic resources are a subset of biological resources.

 | DOI 10.1163/9789004360273_006

in UNCLOS, neither in Part XI concerning the Area and its resources, nor in Part XII dealing with the protection of the environment, nor in Part XIII on scientific research and finally not in Part VII on the High seas. The focus on genetic resources and the benefits possibly derived therefrom and the necessity for a management scheme for the sustainable utilization of genetic resources only became an issue under the Biodiversity Convention.

In dealing with marine genetic resources one has to distinguish clearly between those belonging to the deep seabed – the Area – and those in the water column above the Area. I shall only deal with these two since they are the ones under political consideration. It is being argued that since the Convention does not deal with genetic resources – at least not adequately – it is necessary to close the existing legal gap. Whether and to what extent such legal gap exists is the topic of controversial discussion.

My working hypothesis is that the management of marine genetic resources in areas beyond national jurisdiction should be based upon the principle of sustainability taking into account the Biodiversity Convention and its Nagoya Protocol. Although this is a concrete issue, it is part and parcel of a broader issue: namely, whether UNCLOS constitutes a fixed legal regime for the oceans, or whether it is open for progressive development. If we follow the latter position that UNCLOS is open – which is in my view the only sustainable answer – the question arises which are the appropriate means of a progressive development of UNCLOS and who has the mandate to initiate and implement such legislative progressive development. These are in my view the Meeting of States Parties as well as States acting together on the global or regional level and the international courts and tribunals which have the jurisdiction concerning the interpretation and application of UNCLOS. However, I hasten to add that there are limits to modifying or further developing UNCLOS by subsequent agreements or interpretation, lest it lose its character as a constitution for the oceans based upon a comprehensive and in many ways balanced approach as emphasized in its Preamble. I shall come back to this point at the end of my presentation.

Having established the theoretical background of my presentation, let me now turn to the more concrete question as to whether and to what extent the principle of sustainable development – which I consider essential for solving the problems concerning marine genetic resources in areas beyond national jurisdiction – already influences the interpretation of UNCLOS or – considering that this principle is the *leitmotiv* of international environmental law – whether particular action by the international community should be taken to make this principle operative beyond what is presently possible under the Convention for the sake of managing the utilization of marine genetic resources adequately.

The term "sustainable management" refers to measures which shall lead to sustainable development. ITLOS in its Advisory Opinion concerning sustainable fishing refers to sustainable management, defining it as 'conservation and development'.[3] It further states that the ultimate goal of sustainable management of fish stocks is 'to develop and conserve them as a viable and sustainable resource'.[4] I consider this definition to be fully acceptable.

The principle that States should ensure the sustainable development of natural resources under their jurisdiction or control is considered to have been coined by the Brundtland Report,[5] defining the principle of sustainable development as 'development that meets the need of the present without compromising the ability of future generations to meet their own need'. From there it influenced the outcome of the UN Conference on Environment and Development, in particular the Declaration of Rio, the Agenda 21 of 1992 and the Convention on Biological Diversity. It is evident that UNCLOS, which was opened for signature ten years earlier in 1982, belongs to the pre-sustainable development period and therefore one looks in vain for any explicit reference to the principle of sustainability in the Convention.

Although there are uncertainties concerning what sustainability means, the principle is generally considered to comprise alongside several procedural aspects four substantive elements:[6] the need to preserve natural resources for the benefit of future generations;[7] the aim of exploiting natural resources in a manner which is rational; the equitable use of natural resources, which means taking into consideration the needs of other States; and finally, the need to ensure that environmental considerations are integrated in development plans or policies.[8] The latter includes amongst others the precautionary principle. These four elements are closely interrelated.

Whilst the Convention was adopted before the principle of sustainability became the *leitmotiv* of international environmental law, it nevertheless foreshadowed and formulated several of the core elements of that principle. Apart from that, its interpretation was influenced by the sustainability principle, as can be seen in the Advisory Opinion of ITLOS. But due to new scientific

3 *Request for an advisory opinion submitted by the Sub-Regional Fisheries Commission (SRFC)*, Advisory Opinion, ITLOS Reports 2015, para. 191.

4 Ibid., para. 190.

5 Report of the World Commission on Environment and Development, 'Our Common Future', annexed to UNGA Res. 42/427 (4 August 1987), p. 43 (Brundtland Report).

6 P. Sands, *Principles of International Environmental Law: Frameworks, standards and implementation*, Manchester University Press, 1994, pp. 198 *et seq.*

7 See, in particular, E.B. Weiss, *In Fairness to Future Generations: International Law, Common Patrimony and Intergenerational Equity*, United Nations University, 1989.

8 Sands, n. 7.

findings, insights and developments the issue concerning the management of marine genetic resources has emerged which may require some action from the international community.

Let me go through the four main elements of sustainable management/ development and test them against UNCLOS.

2 For Future Generations

The idea that natural resources are managed in a way that they may also benefit future generations is well established in international law and continues to gain ground. Already in the Pacific Fur Seals Arbitration (1893) it was emphasized that the slaughtering of nursing female seals was deteriorating the stock of Pacific fur seals and, therefore, should not be tolerated.[9] Recent international agreements, in particular the Biodiversity Convention, follow the same approach. It is generally agreed that the concept of intergenerational equity is inherent in the common heritage principle (UNCLOS, Art. 136). The consequences which flow from this principle are detailed in Part XI of the Convention. They comprise the sharing of benefits (Article 140), the transfer of technology (Article 144), accommodation of activities in the deep sea area and in the marine environment (Article 147), the protection of the marine environment (Art. 145) and the equal participation of all States in deep seabed activities (Art. 148). The most striking provisions in this respect are those on the accommodation of deep seabed mineral resource activities to the need of competing terrestrial mining activities (Art. 150).[10] It is to be expected that the Seabed Authority will develop and implement such policy goals when exploitation begins.

3 Rational Use

The idea that natural resources – renewable as well as non-renewable ones – are to be used rationally has been highlighted in several international agreements. International law has undergone a significant development in this respect. This element has been clearly recognized by the Convention read

9 Award between the United States and the United Kingdom relating to the rights of jurisdiction of United States in the Bering's sea and the preservation of fur seals, RIAA, vol. XXVIII, 1893, pp. 263–276.

10 These provisions were highly controversial.

together with the Straddling Fish Stocks Agreement. Rational use means that the resource in question must not been used to an extent or in a way that threatens the existence and/or the sustainability of the resource in question or dependent or associated resources. All forms of utilization have to be undertaken in a way that honours these considerations. There exists quite some jurisprudence affirming this aspect of sustainable development.

4 Equity in International Environmental Affairs

The third aspect of the principle of sustainable management, namely the application of equity in international environmental affairs, even predates the United Nations Conference on Environment and Development of 1992. It originally focused on the allocation of natural resources and, in particular, on financial and other contributions to the protection of the environment.[11]

Two different scenarios are broadly governed by this aspect: first, when it comes to taking or financing conservation and restoration measures, the need to take into consideration what caused the situation which made such measures necessary in the first place; and second, the participation in and contribution to international environmental organizations or arrangements. Perhaps the most prominent example concerning the implementation of equity in sharing resources is to be found in Part XI of the Convention and Annex III and IV thereto establishing the legal regime on deep seabed mining. It would exceed the scope of this presentation, but it is safe to say that object and purpose of that regime is to guarantee the equitable use of the resources in question.[12]

5 Integration of Environmental Considerations in Economic Development Plans and Policies

This fourth aspect of the principle of sustainable management is of a mostly procedural nature and has several facets, some of which are firmly established in international treaties or international proclamations. This aspect is reflected most comprehensively in Principle 4 of the Rio Declaration of 1992 which states: 'In order to achieve sustainable development, environmental protection shall constitute an integral part of the development process and cannot

11 Sands, n. 7, p. 204.

12 See amongst others R. Wolfrum, *Internationalisierung staatsfreier Räume*, Springer, 1984.

be considered in isolation from it'. The full implementation of this approach on the international level, in particular in the context of WTO disputes, would have significant consequences. Art. XX lit. g GATT 1994, which was historically meant for resources such as oil and gas, offers at least in theory such a possibility, providing that States may take measures to protect exhaustible natural resources. In accordance with a more dynamic interpretation of the GATT, the Appellate Body in *US – Shrimp* found that the generic term "natural resources" was not static, but had to take into account the relevance attached to living resources as an environmental resource.[13] The principle of the interrelationship of resource management with social and economic factors is further clearly established in Art. 5 of the Straddling Fish Stocks Agreement. The application of this approach requires the States concerned to weigh the various interests, in particular with a view to guaranteeing a sustainable development of the resource concerned. This is a complex undertaking, particularly if other States for whatever reason put the emphasis differently. In the hearing in the procedure concerning the Advisory Opinion of ITLOS, the issue was brought up whether the objective of producing fish for human consumption should be given priority over fishing with a view to producing fertilizer. This question is by no means unique in the maritime context.

6 Problems of Realizing the Principle of Sustainable Management/Development in Respect of Marine Genetic Resources Beyond Areas under National Jurisdiction

Although the regimes concerning the management and conservation of fish resources established under UNCLOS and the Straddling Stocks Agreement provide for the realization of the sustainability principle, neither of them is tailored to a sustainable development/management of marine genetic resources. Equally the legal regime concerning deep seabed mining is not tailored adequately to deal with marine genetic resources in the Area. The Area is described as 'the seabed and ocean floor and subsoil thereof beyond the limits of national jurisdiction'. This excludes *de lege lata* the water column above the Area (Art. 1(1)(1) UNCLOS). This limitation to the internationalization of the high seas was intended. Part XI of the Convention applies in general only to 'activities in the Area' which is defined in Art. 1(1)(3) of the Convention as 'all activities of exploration for and exploitation of , the resources of the Area'

13 Appelate Body Report, *United States – Import Prohibition of Certain Shrimp and Shrimp Products*, adopted 6 November 1998, WT/DS58/AB/R, para. 130.

while resources are defined in Art. 133(a) of the Convention as 'all solid, liquid, or gaseous mineral resources *in situ* in the Area at or beneath the seabed, including polymetallic nodules'. This limits the deep seabed mining regime in several ways. First, this mining regime is tailored to non-living resources and accordingly does not cover living resources, including genetic resources. However, this does not mean that the Authority does not have the jurisdiction to protect marine genetic resources against negative impacts from mining. Nevertheless, this does not amount to a competence to establish a legal regime on the sustainable utilization of such resources *de lege lata*.

Second, the mining regime does not cover the water column above the high seas and accordingly a regime concerning the sustainable management and development of marine genetic resources cannot, *de lege lata*, be established under Part XI of the Convent.

However, one should not stop at this point. According to Art. 136 UNCLOS, the Area and its resources are the common heritage of mankind. It is to be noted that the Area and the resources are referred to separately, as was also the case in the Declaration of Principles. The Area is dealt with individually and without reference to resources or activities in Art. 138, 141, 143 and 149. This confirms my point that the scope of the legal regimes concerning the Area and concerning activities differ. The underlying motive for such differentiation can be deduced from the legislative history of Part XI. Focussing on activities in the Area was meant to limit the jurisdiction of the Authority. On that basis, it is safe to argue that, due to the declaration of the Area as the common heritage of mankind, the scope of this principle goes beyond the scope of the resource-oriented functions of the Authority. This means the common heritage principle guides actions[14] in the Area not governed by parts of Section 2 and Section 3 et seq. of Part XI. This includes as a matter of logic the management of marine genetic resources *in situ* of the seabed of the Area or the subsoil thereof. Certainly, the application of the common heritage principle to genetic resources does not amount to a fully established legal regime such as the one for mineral resources. Having already established that the principle of sustainable management/development has to be used to identify the consequences of the common heritage principle, elaborating such a legal regime will clearly be an act based upon pre-existing principles of the Convention. Accordingly, the elaboration of the necessary rules is of a different nature than the one dealing with genetic resources of the high seas.

As already indicated, genetic resources of the high seas are not covered by Part XI of the Convention; nor are they addressed in Parts VI, XII, or XIII of

14 To be distinguished from activities as defined in Article 1(3).

the Convention. However, this does not mean that the utilization of marine genetic resources of the high seas is prohibited. Even if not directly listed in Art. 87 of the Convention seas a freedom of the high seas, this does not exclude activities in respect of such marine genetic resources to the extent that the general limits on high seas activities set out in Art. 87 (2) are respected.

I am aware of the problem of this dualistic approach. It may not be feasible to distinguish clearly between genetic resources of the seabed and those of the water column. Apart from that, account has to be taken of coastal States' rights concerning the extended continental shelf. As a matter of practicability, it will be necessary to establish a legal regime which respects the already existing rules and vested rights. But the principle of sustainability should be a neutral starting point.

7 Conclusion

Let me briefly conclude and also come back to the general query raised at the beginning of my presentation:

It is from the point of view of law necessary to differentiate between genetic resources belonging to the international seabed and subsoil thereof and marine genetic resources in the high seas. Whereas the former are covered by the common heritage principle the latter are not. Account has to be taken in this context of coastal States' rights concerning the extended continental shelf.

Establishing a legal regime which spells out the consequences of the common heritage principle in respect of marine genetic resources belonging to the seabed would be an act of interpretation or progressive development of the Convention, whereas the development of a regime on the marine genetic resources of the high seas requires the establishment of a new international instrument. The principle of sustainable management/development would necessarily be the guiding line for such a regime.

From the point of view of safeguarding the coherency of the Convention it is important to respect and implement fully Art. 22 of the Biodiversity Convention in connection with Art. 311 UNCLOS. These articles call for the consistency in the implementation of the Biodiversity Convention and the Convention as well as for the superiority of the Convention in case there is a conflict between the two.

I am aware that the issue I have addressed is the topic of controversial discussion. My contribution to the ongoing discussion is to accept the guidance provided by the principle of sustainability and at the same time protect the integrity of the Convention as the constitution for ocean affairs.

Bibliography

Sands, P., *Principles of International Environmental Law: Frameworks, standards and implementation*, Manchester University Press, 1994.

Weiss, E.B., *In Fairness to Future Generations: International Law, Common Patrimony and Intergenerational Equity*, United Nations University, 1989.

Wolfrum, R., *Internationalisierung staatsfreier Räume*, Springer, 1984.

CHAPTER 6

Toward Sustainable Management of Marine Natural Resources

Yoshifumi Tanaka

1 Introduction

Judge Wolfrum, in his keynote speech given at the Third Hamburg International Environmental Law Conference in 2016, discussed a wide range of issues concerning the sustainable management of both living and non-living resources in the oceans.[1] It was my honour and privilege that I was given an opportunity to make comments on the learned Judge's speech at the conference. Since detailed examination of each and every issue on this subject cannot be made here, I wish to highlight four issues: the concept of sustainable development, consideration of time elements, the interplay between law and science, and *locus standi* in response to breaches of obligations *erga omnes partes*.

2 The Concept of Sustainable Development Revisited

2.1 *Development of the Concept of Sustainable Development*

Sustainable development is a key concept in the management of marine natural resources. Thus, first of all, it is necessary to examine the concept of sustainable development. This concept was given currency by the Report of the World Commission on Environment and Development (WCED), 'Our Common Future'. In its Report, the Commission defined this concept as 'development that meets the needs of the present without compromising the ability of future

* This contribution is based on the text of oral comments made by the author on 15 April 2016 at the HIELC 2016.

1 R. Wolfrum, 'Sustainable Management of Ocean Ecosystems', 3rd Hamburg International Environmental Law Conference, 15 April 2016, at Bucerius Law School, Hamburg, Germany. Although there is no generally agreed definition of 'sustainable management', it can be considered as management to achieve sustainable development.

 | DOI 10.1163/9789004360273_007

generations to meet their own needs'.[2] At present, the concept of sustainable development is increasingly incorporated into treaties and other instruments relating to the conservation of marine living resources.[3] For example, Chapter 17 of Agenda 21 of 1992 provides: 'States commit themselves to the conservation and sustainable use of marine living resources on the high seas'.[4] Likewise the FAO Code of Conduct for Responsible Fisheries recognizes that 'long-term sustainable use of fisheries resources is the overriding objective of conservation and management'.[5] At the treaty level, the 1995 UN Fish Stocks Agreement clearly states in Article 2: 'The objective of this Agreement is to ensure the long-term conservation and sustainable use of straddling fish stocks and highly migratory fish stocks through effective implementation of the relevant provisions of the Convention'.[6] Similarly, Article 1 of the 1992 Convention on Biological Diversity,[7] which is applicable to marine biological diversity, also makes clear that one of its objectives is the sustainable use of components of biological diversity. Here 'sustainable use' can be regarded as a variant of the concept of sustainable development.

As the International Court of Justice (ICJ) observed in the *Gabčíkovo-Nagymaros Project* case,[8] the concept of sustainable development seeks in essence to reconcile the need for development with environmental protection. Given that marine living resources are renewable resources, sustainable use of these resources is of central importance. Furthermore, as Judge Wolfrum suggested,[9] the concept of sustainable development also applies to exploita-

2 World Commission on Environment and Development, *Our Common Future*, Oxford University Press, 1987, p. 43.

3 See N. Schrijver, *The Evolution of Sustainable Development*, Nijhoff, 2008, pp. 102 *et seq.*

4 Agenda 21, para. 17.46, available at: <https://sustainabledevelopment.un.org/content/documents/Agenda21.pdf> accessed September 2017.

5 Article 7.2.1. This document was unanimously adopted on 31 October 1995 by the FAO Conference. For the text, see <http://www.fao.org/fi/agreem/codecond/ficonde.asp> accessed September 2017.

6 1995 Agreement for the Implementation of the Provisions of the United Nations Convention on the Law of the Sea of 10 December 1982 Relating to the Conservation and Management of Straddling Fish Stocks and Highly Migratory Fish Stocks (entered into force 11 December 2001) 2167 UNTS 3.

7 Entered into force 29 December 1993, available at https://www.cbd.int/convention/text/ accessed September 2017.

8 *Gabčíkovo-Nagymaros Project* (Hungary v. Slovakia), Judgment, ICJ Reports 1997, p. 78, para. 140. See also The *Arbitration regarding the Iron Rhine* ("Ijzeren Rijn") *Railway* case (Belgium v. Netherlands), 27 RIAA pp. 28–29, para. 59.

9 Wolfrum, n. 1.

tion of non-living resources since these resources must be exploited without compromising the ability to meet the needs of future generations. All in all, one can argue that the concept of sustainable development provides a guiding principle when deciding policy and measures with regard to the management of marine natural resources.

2.2 *Normative Status of the Concept of Sustainable Development*

However, the concept of sustainable development leaves two issues that need further consideration with regard to its normative status. The first issue concerns the exact legal substance of the concept of sustainable development.[10] In this regard, Judge Wolfrum, highlighted four elements of the concept:[11]

- the need to preserve natural resources for the benefit of future generations,
- the aim of exploiting natural resources in a manner which is rational,
- the equitable use of natural resources, which means taking into consideration the needs of other States, and
- the need to ensure that environmental considerations are integrated in development plans or policies.

It could well be said that these elements are regarded as core components of the concept of sustainable development.

However, this does not mean that there is uniform understanding with regard to the substance of this concept. For instance, some commentators consider that procedural elements, such as environmental impact assessments and public participation in decision-making, also constitute components of the concept of sustainable development.[12] In fact, the New Delhi Declaration of Principles of International Law Relating to Sustainable Development, adopted on 2 April 2002 by the International Law Association, states that public participation is essential to sustainable development.[13] This point was subsequently

10 Schrijver, n. 3, pp. 221–222; Y. Tanaka, *A Dual Approach to Ocean Governance: The Cases of Zonal and Integrated management in International Law of the Sea*, Ashgate, 2008, pp. 71–73.

11 Wolfrum, n. 1.

12 A.E. Boyle and D. Freestone, 'Introduction' in A.E. Boyle and D. Freestone (eds.), *International Law and Sustainable Development, Past Achievements and Future Challenges*, Oxford University Press, 1999, pp. 8–16.

13 See para. 5.1. The Declaration was reproduced in (2002) 2 *International Environmental Agreements: Politics, Law and Economics*, pp. 211–216.

confirmed by the 2012 Sofia Guiding Statements.[14] Furthermore, the normative status of each component is a matter for debate. Inter-relationship between the components is also less clear. Moreover, the concept of sustainable development is not static and its contents may increase over time.[15] Overall it may have to be admitted that there is considerable uncertainty as to the normative substance of the concept of sustainable development. Hence some doubts could be expressed regarding whether and to what extent this concept can legally constrain the behaviour of States.

The second issue relates to the justiciability of the concept of sustainable development.[16] The concept of sustainable development ultimately requires a change in the quality and patterns of life. In the words of the WCED Report,

> In essence, sustainable development is *a process of change* in which the exploitation of resources, the direction of investments, the orientation of technological development, and institutional change are all in harmony and enhance both current and future potential to meet human needs and aspirations [emphasis added].[17]

Similarly, Principle 8 of the Rio Declaration states that: 'To achieve sustainable development and a higher quality of life for all people, States should reduce and eliminate unsustainable patterns of production and consumption and promote appropriate demographic policies'. As shown in these statements, sustainable development is closely interlinked to the national policy of a State. Hence it may be less easy for international courts and tribunals to determine whether a State's conduct is contrary to sustainable development in international law. Thus it is open to debate whether the concept of sustainable development itself can be an independent rule for adjudication.[18] In fact, the ICJ in the *Gabčíkovo-Nagymaros Project* case appeared to regard sustainable development as only a 'concept', not a principle or a rule.[19] All in all, the normativity of

14 2012 Sofia Guiding Statements on the Judicial Elaboration of the 2002 New Delhi Declaration of Principles of International Law Relating to Sustainable Development, para. 7, available at: http://www.ila-hq.org/index.php/committees accessed September 2017.

15 Schrijver, n. 3, p. 208.

16 Tanaka, n. 10, pp. 73–75.

17 WCED, n. 2, p. 46.

18 Boyle and Freestone, n. 12, p. 16; A.V. Lowe, 'Sustainable Development and Unsustainable Arguments', in Boyle and Freestone, n. 12, pp. 24–25.

19 P. Sands, 'International Courts and the Application of the Concept of "Sustainable Development"', *Max Planck Yearbook of United Nations Law*, vol. 3, 1999, p. 393.

the concept of sustainable development remains modest as a rule of conduct and a rule for adjudication. Hence the function of the concept of sustainable development must be sought in a different area of law.

2.3 *Interstitial Function of the Concept of Sustainable Development*

The role of norms is not limited to the function as rules of conduct and/or rules for adjudication. Norms also perform an interstitial function to adjust overlapping or conflicting norms. It is in this area of norms that the normative force of the concept of sustainable development should be sought. In this regard, Lowe argues that 'sustainable development enter[s] into the process only after it has been determined that the overlapping or conflicting primary norms are applicable'.[20] In his view, sustainable development is 'a legal concept exercising a kind of interstitial normativity, pushing and pulling the boundaries of true primary norms when they threaten to overlap or conflict with each other'.[21] Faced with competing norms, international courts and tribunals may take this concept into account in making a judicial choice to lead to the preferred outcome. To this extent, the concept of sustainable development can be used as a factor to be taken into account with a view to ensuring a preferable interpretation. Thus the concept of sustainable development can perform an interstitial function in the judicial process.[22]

An example in this regard is provided by *US – Shrimp* (1998) before the WTO Appellate Body. The preamble of the WTO Agreement explicitly acknowledges 'the objective of sustainable development'. In this regard, the Appellate Body stated that:

> As this preambular language reflects the intentions of negotiators of the *WTO Agreement*, we believe it must add colour, texture and shading to our interpretation of the agreements annexed to the *WTO Agreement*, in this case, the GATT 1994. We have already observed that Article XX(g) of the GATT 1994 is appropriately read with the perspective embodied in the above preamble.[23]

Here the Appellate Body took account of sustainable development reflected in the preamble of the WTO agreement when interpreting Article XX(g), even

---

20 Lowe, n. 18, p. 31.

21 Ibid.

22 Ibid., pp. 34–35.

23 Appellate Body Report, *United States – Import Prohibition of Certain Shrimp and Shrimp Products*, WT/DS58/AB/R, adopted 6 November 1998, para. 153.

though it eventually concluded that the United States measure fails to meet the requirements of the chapeau of Article XX and therefore is not justified under Article XX of the GATT 1994.[24] In summary, if sustainable development cannot be regarded as an independent rule for adjudication, it can nevertheless be taken into account as a judicial factor when interpreting rules governing the management of marine natural resources, including relevant provisions of the UN Convention on the Law of the Sea (UNCLOS),[25] in the process of judicial reasoning.[26]

3 Consideration of Time Elements

3.1 *Inter-Temporal Nature of Sustainable Development*

Sustainability concerns longevity and persistency in the most fundamental sense. In this sense, the concept of sustainable development itself contains time elements.[27] Consideration of time elements is of particular importance in sustainable management of marine natural resources. In this regard, three types of time elements must be taken into account: time in the ecological sense, time in the precautionary sense, and time in an inter-generational sense.

First, ecosystems need a certain period of time to recover and regenerate from exploitation or environmental damage.[28] Accordingly, there is a need to take account the period of time necessary to restore marine ecosystems in sustainable management of marine living resources. The time element in this ecological sense is reflected in UNCLOS, Article 61(3): 'Such measures shall also be designed to maintain or *restore populations of harvested species* at levels which can produce the maximum sustainable yield [emphasis added]'. Likewise UNCLOS, Article 119(1)(a) also stipulates that:

> In determining the allowable catch and establishing other conservation measures for the living resources in the high seas, States shall:

24 Ibid., para. 187(c).

25 United Nations Convention on the Law of the Sea (adopted 10 December 1982, entered into force 16 November 1994) 21 ILM 1261 (1982) (UNCLOS).

26 In this regard, Sofia Guiding Statements explicitly state that: 'Treaties and rules of customary international law should be interpreted in the light of principles of sustainable development'. Sofia Guiding Statements, supra note 12, para. 2.

27 Schrijver, n. 3, pp. 208–209; Separate Opinion of Judge Cançado Trindade in the *Case Concerning Pulp Mills on the River Uruguay* (Argentina v. Uruguay), Separate Opinion, ICJ Reports 2010, p. 185, para. 133.

28 Schrijver, n. 3, p. 213.

> (a) take measures which are designed, on the best scientific evidence available to the States concerned, to maintain or *restore populations of harvested species* at levels which can produce the maximum sustainable yield [emphasis added].

The passage of time is also needed to allow ecosystems to adapt to environmental change. This point is reflected in Article 2 of the United Nations Framework Convention on Climate Change (UNFCCC),[29] which refers to 'a time frame sufficient to allow ecosystems to adapt naturally to climate change, to ensure that food production is not threatened and to enable economic development to proceed in a sustainable manner'. To fully incorporate time elements in the ecological sense, the ecosystem approach must be applied to the management of marine living resources. Hence the ecosystem approach and sustainable development are closely inter-linked.[30] Once mineral resources have been exploited in the oceans, they are not normally restored. In appropriate circumstances, however, there will be a need to take account of the period of time that is necessary to restore damaged ecosystems surrounding an exploitation area. In this sense, time elements in the ecological sense must also be considered in the exploitation of non-living resources.

Second, in light of growing concern over the depletion of marine living resources, the precautionary approach has increasingly become enshrined in international instruments concerning the conservation of marine living resources. In this regard, UN General Assembly Resolution of 8 December 2015 recognizes 'the urgent need for action at all levels, relying on scientific advice, to ensure the long-term sustainable use and management of fisheries resources through the wide application of the precautionary approach and ecosystem approaches'.[31] In light of the scientific uncertainty relating to the mechanisms of marine ecosystems, it is logical that the precautionary approach must be applied to the sustainable management of marine living resources.[32] The precautionary approach aims to take preventive measures in order to respond to probable or potential risks which cannot be objectively identified through

29 Entered into force 21 March 1994. The English text is available at: <http://unfccc.int/files/essential_background/background_publications_htmlpdf/application/pdf/conveng.pdf> accessed September 2017.

30 For the ecosystem approach, see Tanaka, n. 10, pp. 75–82.

31 UNGA Res. A/RES/70/75 (8 December 2015), preambular para. 15.

32 Dupuy considers the precautionary approach as one of the elements of sustainable development. P-M Dupuy, 'Où en est le droit international de l'environnement à la fin du siècle', *RGDIP*, vol. 101, 1997, pp. 888–891.

present-day science but which might create environmental damage in the future. In this sense, the precautionary approach is inter-temporal by nature. It can be considered, therefore, that a time element in the precautionary sense is incorporated into the sustainable management of marine natural resources through the application of the precautionary approach.

Third, the concept of sustainable development is inter-temporal in the sense that it seeks to safeguard the needs of present and future generations.[33] As noted earlier, the WCED explicitly refers to 'future generations' as part of the definition of sustainable development. The WCED further stated that: 'Sustainable development seeks to meet the needs and aspirations of the present without compromising the ability to meet those of the future'.[34] Therefore, the concept of sustainable development can be seen as a reflection of 'intergenerational equity'.[35]

3.2 *Evolutionary Treaty Interpretation*

The consideration of time elements is also at issue when interpreting treaty provisions concerning the sustainable management of marine natural resources. Once a treaty is concluded, its text is frozen in time, unless it is amended. However, the political, economic, ecological and technological conditions which constituted the basis for a treaty may change with time. In fact, sustainability of marine resources may change over time since the oceans are dynamic by nature. Furthermore, environmental knowledge and technology are developing rapidly. Such new developments must be reflected in the interpretation and application of treaties in this field. An issue thus arises as to how one can take account of the change and development in the interpretation and application of these treaties. Here evolutionary treaty interpretation comes into play.[36]

According to evolutionary treaty interpretation, treaties are considered as a 'living instrument', and the texts of the treaties are to be interpreted in an

---

33 Schrijver, n. 3, p. 214; Y. Matsui, 'The Road to Sustainable Development: Evolution of the Concept of Development in the UN' in K Ginther et al. (eds.), *Sustainable Development and Good Governance*, Kluwer, 1995, p. 69.

34 WCED, n. 2, p. 40.

35 New Delhi Declaration, n. 13, para. 2.1.

36 Generally on this issue, see M. Fitzmaurice, 'Dynamic (Evolutive) Interpretation of Treaties: Part I', *Hague Yearbook of International Law*, vol. 21, 2008, p. 101; by the same author, 'Dynamic (Evolutive) Interpretation of Treaties: Part II', *Hague Yearbook of International Law*, vol 22, 2009, p. 3. For a recent study on this subject, see C. Djeffal, *Static and Evolutive Treaty Interpretaiton: A Functional Reconstruction*, Cambridge University Press, 2016.

evolutionary manner by taking the development of norms and present-day standards into account. With prudence and caution, evolutionary interpretation may be relevant in the interpretation of provisions of treaties with regard to sustainable management of marine natural resources. An issue to be examined in this context concerns conditions that would justify evolutionary treaty interpretation. As discussed elsewhere,[37] three elements merit highlighting.

The first element is the use of the generic term. According to Judge Higgins, a 'generic term' is 'a known legal term, whose content the Parties expected would change through time'.[38] The evolutionary nature of the generic term has been affirmed by various international courts and tribunals. For instance, the arbitral tribunal, in the 2005 *Iron Rhine Railway* case between Belgium and the Netherlands, held that:

> It has long been established that the understanding of conceptual or generic terms in a treaty may be seen as 'an essentially relative question; it depends upon the development of international relations' (*Nationality Decrees Issued in Tunis and Morocco, P.C.I.J. Series B, No. 4* (1923), p. 24).[39]

Furthermore, in the *United States-Import Prohibition of Certain Shrimp and Shrimp Products*, the Appellate Body of WTO took the view that:

> [T]he generic term 'natural resources' in Article XX(g) [of the WTO Agreement] is not 'static' in its content or reference but is rather 'by definition, evolutionary'.[40]

Thus, when the Appellate Body interpreted the term 'natural resources' in Article XX(g) of the WTO Agreement, it held that measures to conserve exhaustive natural resources, whether living or non-living, may fall within that provision.[41] The Appellate Body's view provides an insight into the interpretation of treaty provisions concerning marine natural resources.

---

37 This issue was discussed by the author in the following article: Y. Tanaka, 'Reflections on Time Elements in the International Law of the Environment', *ZaöRV /Heidelberg Journal of International Law*, vol. 73, 2013, pp. 139–175.

38 Declaration of Judge Higgins in *Kasikili/Sedudu Island* (Botswana v. Namibia), ICJ Reports 1999, 1113, para. 2.

39 *Arbitration regarding the Iron Rhine Railway*, n. 8, p. 73, para. 79.

40 United States – Import Prohibition of Certain Shrimp and Shrimp Products, WT/DS58/AB/R, 12 October 1998, p. 48, para. 130.

41 Ibid., p. 50, para. 131.

The second element is the object and purpose of the treaty referred to in Article 31(1) of the 1969 Vienna Convention on the Law of Treaties (hereafter the Vienna Convention).[42] As Higgins pointedly observed, 'intention of the parties is often to be deduced from the object and purpose of the agreement'.[43] To a certain extent at least, the object and purpose of treaties provide guidance for determining whether or not the parties to the treaty were thought to have committed themselves to a programme of progressive development. As noted, the sustainable management of marine natural resources requires inter-temporal consideration. Hence it can be reasonably presumed that a treaty whose object is to achieve the sustainable management of marine natural resources must by its very nature include an evolutionary content.

The third element is Article 31(3)(c) of the Vienna Convention which requires '(c) any relevant rules of international law applicable in the relations between the parties' to be taken into account in treaty interpretation.[44] Some international decisions seem to support the use of this provision in taking account of inter-temporal elements in treaty interpretation. By referring to this provision, the arbitral tribunal in the OSPAR *Arbitration* case between the United Kingdom and Ireland ruled that:

> Lest it produce anachronistic results that are inconsistent with current international law, a tribunal must certainly engage in *actualisation* or contemporization when construing an international instrument that was concluded in an earlier period.[45]

Likewise, by referring to Article 31(3)(c) of the Vienna Convention, the arbitral tribunal in the *Iron Rhine Railway* case held that provisions of general international law are applicable to the relations between the Parties, and thus should be taken into account in interpreting Article XII of the 1839 Treaty of Separation and Article IV of the Iron Rhine Treaty. In this regard, the arbitral tribunal considered that environmental law and the law on development stand not as

42 1155 UNTS 331. Entered into force 27 January 1980.

43 R. Higgins, 'Time and the Law: International Perspectives on an Old Problem', *International and Comparative Law Quarterly* (*ICLQ*), vol. 46, 1997, p. 519.

44 For a detailed analysis of Article 31(3)(c) of the Vienna Convention, see C. McLachlan, 'The Principle of Systemic Integration and Article 31(3)(c) of the Vienna Convention', *ICLQ*, vol. 54, 2005, pp. 279–320; D. French, 'Treaty Interpretation and the Incorporation of Extraneous Legal Rules', *ICLQ*, vol. 55, 2006, pp. 281–314.

45 Emphasis original. *Dispute Concerning Access to Information under Article 9 of the OSPAR Convention* (Ireland v. United Kingdom), OSPAR Arbitration of 2 July 2003, p. 34, para. 103.

alternatives but as mutually reinforcing, integral concepts, which require that where development may cause significant harm to the environment there is a duty to prevent, or at least mitigate, such harm. In the view of the tribunal, this duty has now become a principle of general international law. Hence this principle applies in activities undertaken in implementation of specific treaties between the Parties.[46] In the *Iron Rhine Railway* arbitration, Article 31(3)(c) opened the way to take the new development of general international law into account in the interpretation of the 1839 Treaty of Separation. In summary, three elements, i.e. the use of a generic term, the inter-temporality reflected in the object and purpose of the treaties, and Article 31(3)(c) of the 1969 Vienna Convention, may provide elements for judging the validity of evolutionary treaty interpretation.

3.3 *Treaty Interpretation in the Whaling in the Antarctic Case*

In considering treaty interpretation concerning marine natural resources, an interesting example is provided by the *Whaling in the Antarctic* case.[47] A central issue in this case was whether JARPA II, Japan's scientific whaling programme, can be regarded as scientific research under Article VIII of the 1946 International Convention for the Regulation of Whaling (ICRW).[48]

The preamble of the ICRW indicates two objectives: conservation of all species of whales and sustainable exploitation of whale stocks to make possible the orderly development of the whaling industry. In this regard, Australia claimed that owing to the evolutionary development of management tools through the incorporation of a Schedule of regulations, 'the international legal regime for the regulation of whaling has evolved from a system primarily designed to manage the exploitation of a natural resource to an increasingly conservation-oriented regime'.[49] According to Australia, International Whaling Commission (IWC) resolutions must be taken into account in the Court's interpretation of Article VIII because they comprise 'subsequent agreement between the parties regarding the interpretation of the treaty' and 'subsequent practice in the application of the treaty which establishes the agreement of the parties regarding its interpretation', within the meaning of subparagraphs (*a*) and (*b*), respectively, of Article 31(3) of the Vienna Convention. By referring to IWC resolutions, Australia asserted that Article VIII(1) authorizes the

46 *The Arbitration Regarding the Iron Rhine Railway*, n. 8, pp. 66–67, paras. 58–59.

47 *Whaling in the Antarctic* (Australia *v.* Japan: New Zealand intervening), Judgment, ICJ Reports 2014, p. 226.

48 Text in: 161 UNTS 74. Entered into force 10 November 1948.

49 Memorial of Australia, 9 May 2011, p. 62, para. 2.125.

granting of special permits to kill whales only when non-lethal methods are not available.[50] It appears that Australia claimed a sort of evolutionary interpretation of the ICRW.

The Court did not admit Australia's argument for two reasons. First, IWC resolutions cannot be regarded as subsequent agreement to an interpretation of Article VIII, nor as subsequent practice establishing an agreement of the parties regarding the interpretation of the treaty within the meaning of subparagraphs (*a*) and (*b*), respectively, of Article 31(3) of the Vienna Convention. Second, the relevant resolutions and Guidelines that have been approved by consensus do not establish a requirement that lethal methods be used only when other methods are not available.[51]

However, it is not suggested that the Court completely denied the evolutionary nature of the ICRW.[52] A key element in this regard is a duty to cooperate with the IWC.[53] The Court ruled that: 'the States parties to the ICRW have a duty to co-operate with the IWC and the Scientific Committee and thus should give due regard to recommendations calling for an assessment of the feasibility of non-lethal alternatives'.[54] It follows that Japan is under an obligation to give due regard to such recommendations.[55] In fact, the Court ruled that little analysis of the feasibility of using non-lethal methods to achieve the JARPA II research objectives is difficult to reconcile with Japan's obligation to give due regard to IWC resolutions and Guidelines.[56] Following the Court's view, the State parties to the ICRW are required to pay due regard to the evolution of the regulation of whaling through IWC resolutions when implementing relevant provisions of the Convention. In so doing, the Court incorporated a time element, i.e. the evolution of whaling regulation, into the interpretation and application of provisions of the ICRW. The duty to cooperate pronounced by the Court is noteworthy since it can open the way to the incorporation of evolutionary elements into treaty interpretation. Under this approach, however,

50 ICJ Reports 2014, p. 256, paras. 78–79.

51 Ibid., p. 257, para. 83. See also p. 251, para. 56.

52 In fact, the Court accepted that the functions conferred on the IWC have made the Convention an 'evolving instrument'. Ibid., p. 247, para. 45.

53 For a detailed analysis of this subject, see T. Sato, 'Evolution of International Organisation in the Whaling Case' (in Japanese) in S. Yanai and S. Murase (eds.), *Putting International Law into Practice: In Memory of Ambassador Ichiro Komatsu*, Shinzansha, 2015, pp. 149–180.

54 ICJ Reports 2014, p. 257, para. 83.

55 Ibid., p. 270, para. 137. See also p. 256, para. 80.

56 Ibid., p. 271, para. 144.

there is the risk that the interpretation and application of the provisions of the ICRW become strongly influenced by majority opinions in the IWC.[57]

4 Interplay between Law and Science

4.1 *The Southern Bluefin Tuna Cases*

The third issue to be considered relates to the interplay between law and science in the sustainable management of marine natural resources. The proper conservation of living resources and biological diversity in the oceans become possible only on the basis of sufficient and credible scientific data relating to marine species and ecosystems. In fact, the assessment of the marine scientific data directly influences the determination of maximum sustainable yield (MSY) and the allocation of total allowable catch (TAC). It is no exaggeration to say that credible scientific data is a foundation of the sustainable management of marine natural resources.[58]

At present, the interplay between law and science is increasingly important in international jurisprudence concerning the conservation of marine living resources. An illustrative example in this matter is provided by the *Southern Bluefin Tuna* cases of 1999. A central factor of the *Southern Bluefin Tuna* dispute concerns the divergence of opinion concerning fish stock data, which prompted the Japan's unilateral fisheries investigation. Even though there was no disagreement between the parties that the stock of southern bluefin tuna was at its historically lowest levels,[59] the parties differed as to whether the scientific data available showed an upward trend from the current level and whether Japan's experimental fishing programme would have a negative impact upon the southern bluefin tuna stock.[60] While Japan considered that the southern bluefin tuna stock was on the road to recovery,[61] Australia and New Zealand maintained that the amount of southern bluefin tuna taken under the experimental fishing programme could endanger the existence of

57 Sato, n. 53, p. 171.

58 Generally on this issue, see Tanaka, n. 10, pp. 209 et seq.

59 *Southern Bluefin Tuna Cases* (*New Zealand v. Japan; Australia v. Japan*), Provisional Measures, Order of 27 August 1999, 38 ILM 1634, para. 71.

60 Request for Provisional Measures by Australia, p. 75, para. 11; Request for Provisional Measures by New Zealand, p. 11, para. 11, available at: <https://www.itlos.org/en/cases/list-of-cases/case-no-3-4/> accessed September 2017.

61 Response of the Government of Japan to Request for Provisional Measures and Counter-Request for Provisional Measures, p. 162, para. 13, available at: <https://www.itlos.org/en/cases/list-of-cases/case-no-3-4/> accessed September 2017.

the stock.[62] Related to this, a particular difficulty was raised with regard to approaches to the assessment of species to be harvested. In the *Southern Bluefin Tuna* case, there were two principal approaches, which produced very different results: a variable square approach which assumes the absence of fish in areas not currently being fished, and a constant square approach which assumes the same number of fish in those areas as in the areas being fished.[63] In stock assessments, the disputing parties gave different weights to the different hypotheses, resulting in wide disparities in stock assessment.[64] The situation was more complex since there was another uncertainty about the age at which the fish mature, with estimates varying between 8 and 12 years.[65]

ITLOS accepted that there was scientific uncertainty regarding measures to be taken to conserve the stock of southern bluefin tuna and that there was no agreement among the parties as to whether the conservation measures taken so far had led to the improvement in the stock of southern bluefin tuna.[66] In such circumstances, 'the parties should in the circumstances act with prudence and caution to ensure that effective conservation measures are taken to prevent serious harm to the stock of southern bluefin tuna'.[67] ITLOS thus held that:

> although the Tribunal cannot conclusively assess the scientific evidence presented by the parties, it finds that measures should be taken as a matter of urgency to preserve the rights of the parties and to avert further deterioration of the southern bluefin tuna stock.[68]

The Tribunal eventually concluded that 'Australia, Japan and New Zealand shall each refrain from conducting an experimental fishing programme involving the taking of a catch of southern bluefin tuna, except with the agreement of the other parties or unless the experimental catch is counted against its annual national allocation'.[69] Here it may not be unreasonable to consider that ITLOS implicitly applied the precautionary approach. In fact, Judge Laing, in

62 *Southern Bluefin Tuna Cases*, n. 59, p. 1634, para. 74.

63 C.E. Foster, 'The "Real Dispute" in the Southern Bluefin Tuna Case: a Science Dispute?,' *International Journal of Marine and Coastal Law*, vol. 16, 2001, p. 588; Response of the Government of Japan, n. 61, p. 164, paras. 18–19.

64 Ibid., para. 19.

65 Foster, n. 63, p. 577.

66 *Southern Bluefin Tuna Cases*, n. 59, p. 1634, para. 79.

67 Ibid., para. 77.

68 Ibid., para. 80.

69 Ibid., p. 1636, operative part of (d).

his Separate Opinion, stated that: 'it becomes evident that the Tribunal has adopted the precautionary approach for the purposes of provisional measures in such a case as the present'.[70] This view was echoed by Judge Shearer, stating that: 'I believe that the measures ordered by the Tribunal are rightly based upon considerations deriving from a precautionary approach'.[71] Where there is scientific uncertainty with regard to the condition of fish stocks, there would appear to be good reasons for ITLOS to apply the precautionary approach to lead to the most desirable outcome for conservation of marine living resources.[72]

4.2 *The Whaling in the Antarctic Case*

The interplay between law and science was also at the heart of the 2014 *Whaling in the Antarctic* case.[73] In this case, the ICJ ascertained whether special permits granted in relation to JARPA II, Japan's scientific whaling programme, fall within the scope of Article VIII, paragraph I of the ICRW. According to the Court, the two elements 'scientific research' and 'for the purpose of scientific research' are cumulative. It follows that if a whaling programme involves scientific research, the killing of whales pursuant to such a programme does not fall within Article VIII unless these activities are 'for purposes of' scientific research.[74] The Court thus took a two-stage approach.

First, the Court examined whether the design and implementation of JARPA II are reasonable in relation to achieving its stated objectives. After having examined the key elements of JARPA II,[75] it found that the JARPA II activities involving the lethal sampling of whales can broadly be characterised as 'scientific research'.[76]

70 Separate Opinion of Judge Laing, ibid., p. 1642, para. 19.

71 Separate Opinion of Judge *ad hoc* Shearer, ibid., p. 1650.

72 Further, see Y. Tanaka, 'Rethinking *Lex Ferenda* in International Adjudication', *German Yearbook of International Law*, vol. 51, 2008, pp. 491–493.

73 Generally on this issue, see M.M. Mbengue, 'Between Law and Science: A Commentary on the Whaling in the Antarctic Case', *Questions of International Law*, vol. 14, 2015, pp. 3–12; S.R. Tully, '"Objective Reasonableness" as a Standard for International Judicial Review', *Journal of International Dispute Settlement*, vol. 6, 2015, pp. 546–567; M.M. Mbengue and R. Das, 'The ICJ's Engagement with Science: To Interpret or not to Interpret?', ibid., pp. 568–577; G. Gros, 'The ICJ's Handling of Science in the Whaling in the Antarctic Case: A Whale of a Case?', ibid., pp. 578–620; L.C. Lima, 'The Evidential Weight of Experts before the ICJ: Reflections on the Whaling in the Antarctic Case', ibid., pp. 621–635.

74 ICJ Reports 2014, p. 255, para. 71.

75 These elements are: research objectives, research period and area, research methods and sample size, effect on whale stocks. Ibid., pp. 264–267, paras. 113–126.

76 Ibid., p. 267, para. 127.

Second, the Court ascertained whether the killing, taking and treating of whales in pursuance of JARPA II is 'for purposes of' scientific research and thus may be authorized by special permits granted under Article VIII(1) of the ICRW. A controversial issue in this regard was the use of lethal methods. The Court did not consider that the use of lethal methods is *per se* unreasonable in the context of JARPA II.[77] However, one question was whether Japan had considered the possibility of making more extensive use of non-lethal methods before launching JARPA II. The Court answered this question in the negative, stating that: 'the papers to which Japan directed it reveal little analysis of the feasibility of using non-lethal methods to achieve the JARPA II research objectives'.[78] Furthermore, the Court considered that sample sizes, i.e. the number of whales of each species to be killed each year, and the launch date for JARPA II were not driven by strictly scientific considerations;[79] and that the target sample sizes in JARPA II were not reasonable in relation to achieving the programme's objectives.[80] It thus concluded that the special permits granted by Japan for the killing, taking and treating of whales in connection with JARPA II were not 'for purposes of scientific research' pursuant to Article VIII(1) of the ICRW.[81] The *Whaling in the Antarctic* judgment calls for two brief comments.

First, in the *Southern Bluefin Tuna* cases, an essential issue concerned the adverse impact of the unilateral implementation of an experimental fishing programme conducted by Japan. In the *Whaling in the Antarctic* case, however, the ICJ did not offer any opinion on possible negative effects of Japan's scientific whaling on whale stocks.[82] Rather, the central issue in this case concerned the validity of Japan's scientific whaling programme *per se*. Hence the ICJ was to ascertain the contents of the scientific programme which are in essence a matter of science. In this sense, the *Whaling in the Antarctic* case is contrasted with the *Southern Bluefin Tuna* cases.[83] The *Whaling in the Antarctic* judgment

77 Ibid., p. 269, para. 135.

78 Ibid., p. 271, para. 144.

79 Ibid., p. 274, para. 156.

80 Ibid., p. 292, para. 224.

81 Ibid., p. 293, para. 227.

82 A. Telesetsky, D.K. Anton, T. Koivurova, 'ICJ's Decision in Australia v. Japan: giving up the Spear or Refining the Scientific Design?', *Ocean Development and International Law*, vol. 45, 2014, p. 333.

83 M. Koyano, 'The Significance of the Whaling Judgment from the Viewpoint of International Administrative Law' (in Japanese), *International Affairs*, vol. 636, 2014, p. 44.

did seem to demonstrate the willingness of the ICJ to determine issues of science.[84]

Second, attention must be paid to the role of experts in the settlement of international disputes involving issues of science, such as disputes concerning environmental protection and conservation of marine living resources.[85] In the *Whaling in the Antarctic* case, both Japan and Australia called experts at the hearings. In the *Whaling in the Antarctic* judgment, the experts' views were often cited by the Court when examining key issues.[86] The Court seemed to attach particular importance to the fact that the experts called by the disputing parties agreed, or to the fact that the opinion of one expert was not contested by a party.[87] For instance, it held that:

> The evidence shows that the JARPA II Research Plan lacks transparency in the reasons for selecting particular sample sizes for individual research items. This is a matter on which the *experts called by the two Parties agreed* [emphasis added].[88]

Likewise the Court considered that: 'the process used to determine the sample size for minke whales lacks transparency, *as the experts called by each of the Parties agreed* [emphasis added]'.[89]

Overall the experts' views did seem to affect the judgment of the ICJ in the *Whaling in the Antarctic* case. Yet, there is no guarantee that the statements of the experts called by the disputing parties are objective and neutral, even though each expert submitted written testimony.[90] Hence caution is advisable

---

84 Telesetsky et al., n. 82, p. 336. However, Judge Owada questioned whether the ICJ is an appropriate forum to examine issues of science. Dissenting Opinion of Judge Owada, ICJ Reports 2014, p. 310, para. 24. See also M. Fitzmaurice, *Whaling and International Law*, Cambridge University Press, 2015, p. 106.

85 Generally on this issue, see D. Peat, 'The Use of Court-Appointed Experts by the International Court of Justice', *British Year Book of International Law*, vol. 84, 2014, pp. 271–303; C.E. Foster, 'New Clothes for the Emperor? Consultation of Experts by the International Court of Justice', *Journal of International Dispute Settlement*, vol. 5, 2014, pp. 139–173.

86 Fitzmaurice, n. 84, pp. 99–102; S. Sakamoto, 'The Scope of the *Whaling in the Antarctic* Judgment: Japanese Perspectives' (in Japanese), *International Affairs*, vol. 636, 2014, pp. 14–15.

87 Lima, n. 73, p. 633.

88 ICJ Reports 2014, p. 283, para. 188.

89 Ibid., p. 292, para. 225.

90 Lima, n. 73, p. 631.

when relying on the views of experts called by the parties.[91] Furthermore, there appears to be some scope to consider the question whether the ICJ should have appointed experts under Article 50 of the ICJ Statute to advise whether JARPA II can be regarded as a programme conducted for 'the purposes of scientific research'.[92] In any case the *Whaling in the Antarctic* case sheds light on the role of experts in disputes involving issues of science.

5 Locus Standi on the Basis of Obligations Erga Omnes Partes

Finally, some consideration must be given to the responsibility arising from the breach of obligations concerning the non-sustainable use of marine natural resources. A particular issue to be examined in this regard is whether or not a State party to a treaty concerning the conservation of marine living resources can submit a dispute involving the alleged breach of treaty obligations by another State party to an international court or tribunal, even though no damage occurred to the applicant State. This issue concerns *locus standi* in response to the breach of obligations *erga omnes partes*. The 2014 *Whaling in the Antarctic* case provides an important insight into this subject.

In this case, Judge Bhandari asked the following question of Australia: 'What injury, if any, has Australia suffered as a result of Japan's alleged breach of the ICRW through JARPA II?'[93] In response, a Counsel for Australia stated that: 'Australia is seeking to uphold its collective interest, an interest it shares with all other parties'.[94] Likewise another Counsel for Australia replied to Judge Bhandari's question, 'Australia, like all the other States parties to the 1946 Convention, has a common interest in maintaining the integrity of the régime deriving from the Convention'[95] and that 'all States parties to that

91 In this regard, the ICJ in the *Pulp Mills* case stated that: 'despite the volume and complexity of the factual information submitted to it, it is the responsibility of the Court, after having given careful consideration to all the evidence placed before it by the Parties, to determine which facts must be considered relevant, to assess their probative value, and to draw conclusions from them as appropriate'. ICJ Reports 2010, p. 72, para. 168.

92 Peat, n. 85, p. 287; Fitzmaurice, n. 84, p. 103. See also Joint dissenting opinion Judges Al-Khasawneh and Simma in the *Pulp Mills* case, ICJ Reports 2010, p. 110, paras. 3–4.

93 Verbatim Record, 3 July 2013, CR 2013/13, p. 73, para. 35.

94 *Whaling in the Antarctic* (Australia v. Japan: new Zealand Intervening), Presentation by Burmester, Verbatim Record, CR 2013/18, 9 July 2013, p. 28, para. 19. See also presentation by Boisson de Chazournes, ibid., pp. 33–34, paras. 18–20.

95 Ibid., p. 23, para. 18 (translation).

Convention have a common interest in each State complying with its obligations under the Convention and the régime deriving from it'.[96] By referring to the *Crime of Genocide* Advisory Opinion, the Counsel stated that: 'the States parties "do not have any interests of their own; they merely have, one and all, a common interest …" in the context of the régime established under the 1946 Convention'.[97] It can be argued, therefore, that Australia referred the dispute to the ICJ on the basis of the alleged breach of obligations *erga omnes partes* by Japan.[98] Notably, the ICJ accepted the *locus standi* of Australia in this case. It may not be unreasonable to argue that the Court, in the *Whaling in the Antarctic* case, accepted *locus standi* in response to the breach of an obligation *erga omnes partes*, even though the ICJ remained silent on the *erga omnes partes* character of obligations set out in the ICRW.[99]

It appears that the *Whaling in the Antarctic judgment* is in line with the *Belgium/Senegal* judgment.[100] The ICJ, in the 2012 *Belgium/Senegal* case, held that any State Party to the 1984 UN Convention against Torture may invoke the responsibility of another State Party with a view to ascertaining the alleged failure to comply with its obligation *erga omnes partes*. It thus concluded that Belgium, as a State Party to the Convention against Torture, had standing to invoke the responsibility of Senegal for the alleged breaches of its obligations before the ICJ under Articles 6(2) and 7(1) of the Convention.[101] The *locus standi* in relation to a dispute concerning compliance with the obligation *erga omnes partes* is also echoed by Article 48(1) of the ILC Draft Articles on State Responsibility[102] which provides that:

> Any State other than an injured State is entitled to invoke the responsibility of another State … if:

96 Ibid., para. 19 (translation).

97 Ibid., para. 20 (translation). See also *Reservations to the Convention on the Prevention and Punishment of the Crime of Genocide, Advisory Opinion,* ICJ Reports 1951, p. 23.

98 Fitzmaurice, n. 84, pp. 109–110.

99 In this regard, it is to be noted that Japan did not dispute *locus standi* of Australia in the *Whaling in the Antarctic* case.

100 Fitzmaurice, n. 84, p. 110.

101 *Questions relating to the Obligation to Prosecute or Extradite (Belgium* v. *Senegal)*, ICJ Reports 2012, paras. 69–70. See also Separate Opinion of Judge Cançado Trindade, ibid., pp. 527–529, paras. 104–108.

102 This instrument is available at: <http://legal.un.org/ilc/texts/instruments/english/commentaries/9_6_2001.pdf> accessed September 2017.

(a) the obligation breached is owed to a group of States including that State, and is established for the protection of a collective interest of the group; or
(b) the obligation breached is owed to the international community as a whole.

In light of this provision, there may be scope to argue that not directly injured States can institute proceedings against a State that allegedly breached obligations *erga omnes partes* before an international court or tribunal.[103] Although further development of case-law is needed to draw more general conclusions, the *Belgium/Senegal* and *Whaling in the Antarctic* cases appear to hint in the direction that the ICJ would accept *locus standi* on the basis of the alleged breach of obligations *erga omnes partes*, if it can establish its jurisdiction. If this is the case, the ICJ would assume the role of advocate for the international community in the protection of a common interest reflected in treaties. This would open the way to protect a common interest through international adjudication.

At the same time, care should be taken in noting that the Court continues to apply a rigid approach to the consensual basis of its jurisdiction.[104] By way of example, the Court, in the *East Timor* case, considered that

> the *erga omnes* character of a norm and the rule of consent to jurisdiction are two different things. Whatever the nature of the obligations invoked, the Court could not rule on the lawfulness of the conduct of a State when its judgment would imply an evaluation of the lawfulness of the conduct of another State which is not a party to the case. Where this is so, the Court cannot act, even if the right in question is a right *erga omnes*.[105]

103 Following Kawasaki, these States which are not directly injured can be called 'not directly injured States' in the sense that they do not personally suffer any kind of damage, material or moral, although their subjective rights are considered to have been infringed. K. Kawasaki, 'The "Injured State" in the International Law of State Responsibility', *Hitotsubashi Journal of Law and Politics*, vol. 28, 2000, p. 22.

104 M. Kawano, 'The Role of Judicial Procedures in the Process of the Pacific Settlement of International Disputes', *RCADI*, vol. 346, 2009, pp. 419–421; p. 424.

105 *East Timor* (Portugal *v.* Australia), Judgment, ICJ Reports 1995, p. 102, para. 29.

In referring to this dictum, the Court, in the *Armed Activities* case, also ruled that

> The same applies to the relationship between peremptory norms of general international law (*jus cogens*) and the establishment of the Court's jurisdiction: the fact that a dispute relates to compliance with a norm having such a character, which is assuredly the case with regard to the prohibition of genocide, cannot of itself provide a basis for the jurisdiction of the Court to entertain that dispute. Under the Court's Statute that jurisdiction is always based on the consent of the parties.[106]

Furthermore, careful consideration must be given to the question of whether the Court, rather than a treaty commission, is the relevant organ to protect common interests reflected in obligations *erga omnes partes*. In this regard an issue might arise as to compatibility between the non-compliance procedure of a treaty and the Court's judgment.[107] There is also a risk that States withdraw from, or make reservations to the optional clause of the ICJ Statute, if international litigation on the basis of obligations *erga omnes partes* increases.[108]

6 Conclusions

As comments on Judge Wolfrum's key note speech, this contribution discussed four issues concerning the sustainable management of marine natural resources: the concept of sustainable development, consideration of time elements, the interplay between law and science, and *locus standi* on the basis of obligations *erga omnes partes*. The principal findings of the above considerations can be summarized in five points.

---

106 *Armed Activities on the Territory of the Congo* (*New Application: 2002*) (Democratic Republic of the Congo *v.* Rwanda), Jurisdiction of the Court and Admissibility of the Application, Judgment, ICJ Reports 2006, p. 32, para. 64.

107 Koyano, supra note 83, p. 53.

108 After the *Whaling in the Antarctic* judgment, the Japanese Government made the reservation to the optional clause in order to preclude 'any dispute arising out of, concerning, or relating to research on, or conservation, management or exploitation of, living resources of the sea' from the jurisdiction of the ICJ. See <http://www.icj-cij.org/en/declarations> accessed September 2017.

First, sustainable development, which essentially seeks to reconcile the need for development with environmental protection, including the conservation of marine natural resources and ecosystems, is a key concept in the management of natural resources in the oceans. At the practical level, the concept of sustainable development provides a key guideline when determining specific management measures in this field. At the adjudicative level, despite the uncertainty of its normative substance, international courts and tribunals can consider the concept of sustainable development as a factor guiding the proper interpretation of relevant rules in the judicial process where there is a conflict between primary norms. In this sense, the concept of sustainable development performs an interstitial function in international adjudication.

Second, since sustainability of marine natural resources may change over time, an issue arises as to how it is possible to incorporate the change and development into the interpretation and application of relevant provisions of treaties governing the management of these resources. Here evolutionary treaty interpretation may come into play. In this regard, the use of a generic term, the inter-temporality reflected in the object and purpose of the treaties, and Article 31(3)(c) of the 1969 Vienna Convention, may provide elements that would justify the application of evolutionary treaty interpretation.

Third, following the dictum of the ICJ in the *Whaling in the Antarctic* case, one can argue that State parties to a treaty are under an obligation to give due regard to resolutions adopted by a treaty commission. Thus, resolutions of a treaty commission may affect the interpretation and application of treaty provisions, even if the resolutions are not legally binding.

Fourth, scientific evidence is a key element in international disputes concerning the conservation of marine living resources. As shown in the *Southern Bluefin Tuna* and *Whaling in the Antarctic* cases, the interplay between law and science is increasingly important in the settlement of disputes concerning the conservation of marine living resources. As exemplified by the *Southern Bluefin Tuna* cases, however, international courts and tribunals may encounter difficulties when evaluating competing scientific evidence submitted by the parties in dispute. Related to this, the role of experts merits particular notice.

Finally, it is notable that the ICJ, in the *Whaling in the Antarctic* case, accepted the *locus standi* of Australia, even though Australia suffered no damage from Japan's scientific whaling programme. The *Whaling in the Antarctic* judgment, along with the *Belgium/Senegal* judgment, provides an important precedent that not directly injured States can institute proceedings before the ICJ on the basis of a common interest of parties to a treaty or obligation *erga omnes partes*.

Bibliography

Boyle, A.E., and Freestone, D., 'Introduction' in Boyle, A.E. and Freestone, D. (eds.), *International Law and Sustainable Development, Past Achievements and Future Challenges*, Oxford University Press, 1999, p. 8.

Djeffal, C., *Static and Evolutive Treaty Interpretation: A Functional Reconstruction*, Cambridge University Press, 2016.

Dupuy, P-M., 'Où en est le droit international de l'environnement à la fin du siècle', *Revue générale de droit international public*, vol. 101, no. 4, 1997, p. 873.

Fitzmaurice, M., 'Dynamic (Evolutive) Interpretation of Treaties: Part I', *Hague Yearbook of International Law*, vol. 21, 2008, p. 101.

Fitzmaurice, M., 'Dynamic (Evolutive) Interpretation of Treaties: Part II', *Hague Yearbook of International Law*, vol. 22, 2009, p. 3.

Fitzmaurice, M., *Whaling and International Law*, Cambridge University Press, 2015.

Foster, C.E., 'The "Real Dispute" in the Southern Bluefin Tuna Case: a Science Dispute?,' *The International Journal of Marine and Coastal Law*, vol. 16, no. 4, 2001, p. 571.

Foster, C.E., 'New Clothes for the Emperor? Consultation of Experts by the International Court of Justice', *Journal of International Dispute Settlement*, vol. 5, no. 1, 2014, p. 139.

French, D., 'Treaty Interpretation and the Incorporation of Extraneous Legal Rules', *International and Comparative Law Quarterly*, vol. 55, no. 2, 2006, p. 281.

Gros, G. 'The ICJ's Handling of Science in the Whaling in the Antarctic Case: A Whale of a Case?', (2015) 6 *Journal of International Dispute Settlement*, vol. 6, no. 3, 2015, p. 578.

Higgins, R., 'Time and the Law: International Perspectives on an Old Problem', *International and Comparative Law Quarterly*, vol. 46, no. 3, 1997, p. 501.

Kawano, M., 'The Role of Judicial Procedures in the Process of the Pacific Settlement of International Disputes', *Recueil des cours de l'academie de droit international*, vol. 346, 2009, p. 9.

Kawasaki, K., 'The "Injured State" in the International Law of State Responsibility', *Hitotsubashi Journal of Law and Politics*, vol. 28, 2000, p. 17.

Koyano, M., 'The Significance of the Whaling Judgment from the Viewpoint of International Administrative Law' (in Japanese), *International Affairs*, no. 636, 2014, p. 43.

Lima, L.C., 'The Evidential Weight of Experts before the ICJ: Reflections on the Whaling in the Antarctic Case', *Journal of International Dispute Settlement*, vol. 6, no. 3, 2015, p. 621.

Lowe, A.V., 'Sustainable Development and Unsustainable Arguments', in Boyle, A.E. and Freestone, D. (eds.), *International Law and Sustainable Development, Past Achievements and Future Challenges*, Oxford University Press, 1999, p. 19.

Matsui, Y., 'The Road to Sustainable Development: Evolution of the Concept of Development in the UN' in Ginther, K. *et al.* (eds.), *Sustainable Development and Good Governance*, Kluwer, 1995, p. 53.

Mbengue, M.M., 'Between Law and Science: A Commentary on the Whaling in the Antarctic Case', *Questions of International Law*, vol. 14, 2015, p. 3.

Mbengue, M.M., and Das, R., 'The ICJ's Engagement with Science: To Interpret or not to Interpret?', *Journal of International Dispute Settlement*, vol. 6, no. 3, 2015, p. 568.

McLachlan, C., 'The Principle of Systemic Integration and Article 31(3)(c) of the Vienna Convention', *International and Comparative Law Quarterly*, vol. 54, no. 2, 2005, p. 279.

Peat, D., 'The Use of Court-Appointed Experts by the International Court of Justice', *British Year Book of International Law*, vol. 84, 2013, p. 271.

Sakamoto, S., 'The Scope of the *Whaling in the Antarctic* Judgment: Japanese Perspectives' (in Japanese), *International Affairs*, no. 636, 2014, p. 6.

Sands, P., 'International Courts and the Application of the Concept of "Sustainable Development"', *Max Planck Yearbook of United Nations Law*, vol. 3, 1999, p. 389.

Sato, T. 'Evolution of International Organisation in the Whaling Case' (in Japanese) in Yanai, S. and Murase, S. (eds.), *Putting International Law into Practice: In Memory of Ambassador Ichiro Komatsu* (Tokyo, Shinzansha, 2015), p. 149.

Schrijver, N., *The Evolution of Sustainable Development*, Nijhoff, 2008.

Tanaka, Y., *A Dual Approach to Ocean Governance: The Cases of Zonal and Integrated Management in International Law of the Sea*, Ashgate, 2008.

Tanaka, Y., 'Reflections on Time Elements in the International Law of the Environment', *ZaöRV /Heidelberg Journal of International Law*, vol. 73, no. 2, 2013, p. 139.

Tanaka, Y., 'Rethinking *Lex Ferenda* in International Adjudication', *German Yearbook of International Law*, vol. 51, 2008, p. 467.

Telesetsky, A., Anton, D.K., and Koivurova, T., 'ICJ's Decision in Australia v. Japan: giving up the Spear or Refining the Scientific Design?', *Ocean Development and International Law*, vol. 45, no. 4, 2014, p. 328.

Tully, S.R., '"Objective Reasonableness" as a Standard for International Judicial Review', *Journal of International Dispute Settlement*, vol. 6, no. 3, 2015, p. 546.

The World Commission on Environment and Development, *Our Common Future*, Oxford University Press, 1987.

CHAPTER 7

Sustainable Management of Ocean Ecosystems: Some Comments

David Freestone

On 25 September 2015, the UN General Assembly by Resolution 70/1 adopted the Sustainable Development Goals (SDGs).[1] The preamble says 'They are integrated and indivisible and balance the three dimensions of sustainable development: the economic, social and environmental.'[2] The SDGs comprise 17 goals and 169 targets that address a wide spread of issues on the development and environment agenda, from eliminating poverty and hunger, to conservation of terrestrial and aquatic systems, to peace and justice and strong institutions, and gender equality.

In particular they contain major and wide ranging commitments to the sustainable management of the Oceans. For example, UN Members commit that by 2025, they will prevent and significantly reduce marine pollution of all kinds; by 2020, sustainably manage and protect marine and coastal ecosystems; minimize and address the impacts of ocean acidification; by 2020, effectively regulate harvesting and end overfishing, illegal, unreported and unregulated fishing and destructive fishing practices and implement science-based management plans; by 2020, conserve at least 10 per cent of coastal and marine areas; by 2020, prohibit certain forms of fisheries subsidies; by 2030, increase the economic benefits to Small Island Developing States and least developed countries; increase scientific knowledge, develop research capacity and transfer marine technology; and enhance the conservation and sustainable use of oceans and their resources by implementing international law as reflected in the 1982 United Nations Convention on the Law of the Sea

* This contribution is based heavily on a previous publication, David Freestone, "Modern Principles of High Seas Governance: The Legal Underpinnings" (2009) 39/1 *International Environmental Policy and Law* pp. 44–49. I am grateful for the Editor's agreement for me to draw on that earlier paper.

1 UN A/RES/70/1/L.1. Text available at <https://unstats.un.org/sdgs/report/2016/> accessed 11 September 2017.

2 Ibid.

 | DOI 10.1163/9789004360273_008

(UNCLOS), which provides the legal framework for the conservation and sustainable use of oceans and their resources."[3]

This major commitment to sustainable development of the oceans is *de facto* a commitment to sustainable management of human activities in the ocean. It comes at a time when the UN General Assembly (UNGA) is about to decide whether to develop a new International Legally Binding Instrument under the 1982 Law of the Sea Convention[4] to address the whole issue of the sustainable use and conservation of biological diversity in areas beyond national jurisdiction.

Areas beyond national jurisdiction, a term which covers the high seas and the deep sea bed, cover nearly 50% of the surface of Planet Earth. Despite the fact that the comprehensive legal regime developed by UNCLOS has been with us for more than thirty years and has been in force for over twenty years, it has not really protected the high seas from unparalleled impacts from new human activities, such as bottom trawling on seamount ecosystems, and from the increased intensity of existing activities, such as huge increases in maritime transportation, pollution from garbage and traditional fishing techniques. Nor has the legal framework that UNCLOS provides been able to keep pace with the need to regulate either the exploitation of valuable new resources that have been discovered in high seas areas – be they highly vulnerable deep ocean fish species, ocean thermal vents with accompanying life forms that can live in ambient temperatures as hot as 300–600°C, or cold seeps – or proposals for geo-engineering activities such as ocean fertilization.[5] Under UNCLOS, coastal States have jurisdiction over living and non-living resources out to 200 nautical miles from their coastal baselines and over continental shelf resources out to the geological limit of their continental shelf. Beyond that point, UNCLOS envisages the International Seabed Authority having jurisdiction, but only over the non-living resources of the seabed. Hence, there is a lacuna for deep sea or seabed living resources.

3 This is a summary of main items of the text, see above for more details. These were further reinforced by the 2017 Ocean Conference Call for Action adopted by UNGA Res. 71/312 'Our ocean, our future: call for action' (6 July 2014), available at <http://www.un.org/ga/search/view_doc.asp?symbol=A/RES/71/312&Lang=E> accessed 11 September 2017.

4 United Nations Convention on the Law of the Sea (adopted 10 December 1982, entered into force 16 November 1994) 21 ILM 1261 (1982) (UNCLOS).

5 See R. Rayfuse, M. Lawrence and K. Gjerde, 'Ocean Fertilization and Climate Change: the Need to Regulate Emerging High Seas Uses', *International Journal of Marine and Coastal Law*, vol. 23, 2008, pp. 297–326.

A number of sectoral activities in the high seas are governed by existing treaty regimes; such as the regulation of vessels and vessel source pollution by the International Maritime Organization, or dumping at sea by the 1972 London Convention, its 1996 Protocol and a network – albeit by no means a comprehensive network – of species and regional fisheries treaties and arrangements as well as by some of the regional seas conventions.[6] Nevertheless, international concern has been growing at the lack of an adequate comprehensive framework for high seas governance and while the international community is beginning to respond, progress has been slow. In 2004, on the recommendation of the UN Informal Consultative Process on the Oceans and the Law of the Sea (UNICPOLOS), the UNGA agreed to establish an Ad Hoc Open-ended Informal Working Group to study issues relating to the conservation and sustainable use of marine biological diversity beyond areas of national jurisdiction. This Working Group held a series of meetings between 2007 and 2015. At its last meeting, 20–23 January 2015, the Working Group decided after protracted debate to recommend to the UNGA that it: 'Decide to develop an international legally binding instrument under the Convention on the conservation and sustainable use of marine biological diversity of areas beyond national jurisdiction'.[7] The recommendations of the Working Group also envisaged the establishment of a preparatory committee, to begin work in 2016 and to report to the UNGA in 2017 with recommendations on a text. The proposed negotiations will address the topics identified in the package agreed by the Working Group in 2011, namely the conservation and sustainable use of marine biodiversity in areas beyond national jurisdiction, in particular, together and as a whole, marine genetic resources, including questions on the sharing of benefits, measures such as area-based management tools, including marine protected areas, environmental impact assessments, capacity building and the transfer of marine technology.[8]

6 See e.g., R. Warner, K. Gjerde and D. Freestone, 'Regional Governance for Fisheries and Biodiversity', in S.M. Garcia, J. Rice, and A.T. Charles (eds.), *Governance for Fisheries and Marine Conservation: interactions and co-evolution*, Wiley-Blackwell, 2014, pp. 211–224.

7 'Recommendations of the Ad Hoc Open-ended Informal Working Group to study issues relating to the conservation and sustainable use of marine biological diversity beyond areas of national jurisdiction', UN Doc A/69/780* 1(3 February 2015). In its Resolution 69/292 of 19 June 2015, the General Assembly decided to develop an international legally binding instrument under the UNCLOS on the conservation and sustainable use of marine biological diversity of areas beyond national jurisdiction.

8 Ibid., para 6. For a more detailed discussion of the issues involved see D. Freestone, 'Governance of Areas beyond National Jurisdiction: An Unfinished Agenda?', in J. Barrett

In Resolution 69/292 of 19 June 2015, the UNGA agreed to the establishment of a Preparatory Commission, which met four times during 2016 and 2017 and recommended to the General Assembly that formal negotiation begin in 2018.[9]

Throughout this process one of the key elements of what we are now calling the ILBI under the Law of the Sea Convention should contain a number of basic principles governing the exploitation of the high seas. As early as 2008, as a constructive contribution to the dialogue to enumerate these principles more explicitly, the International Union for Conservation of Nature (IUCN) expert group process proposed the articulation of some basic principles of modern ocean governance,[10] and were supported by a number of other international expert groups.[11] The IUCN marine programme took it upon itself to help clarify these existing principles and in September 2008 issued a document: *10 Principles of High Seas Governance* for public review and comment. At the 2008 IUCN 4th World Conservation Congress in Barcelona on October 7, IUCN President Vali Moosa of South Africa chaired a plenary session presenting the IUCN "Ten Principles of High Seas Governance." I was asked to present those principles at that plenary and what follows is an attempt to provide the legal underpinnings for those principles and to update them in the light of subsequent developments.

It should be stressed that these are not new principles – as I seek to demonstrate below all of them are derived from existing regional or global instruments accepted by consensus. However they have never been collected together before in this way and they all require much more rigorous implementation than they are currently receiving.

---

and R. Barnes, *Law of the Sea: UNCLOS as a Living Treaty*, London: British Institute of International and Comparative Law, 2016, pp. 231–266.

9 On the possible content see the Chair streamlined 'non paper' on the possible elements of a possible ILBI at <http://www.un.org/depts/los/biodiversity/prepcom_files/Chairs_streamlined_non-paper_to_delegations.pdf> accessed 11 September 2017.

10 R. Rayfuse et al., *Co-Chairs' Report of Workshop on High Seas Governance for the 21st Century*, New York City, 2007, available at <http://www.iucn.org/themes/marine/pdf/IUCN%20Workshop%20Co-Chairs%20Summary_FINAL%20VERSION.pdf> accessed 11 September 2017.

11 See: B. Cicin-Sain and D. Freestone, *Report from the Strategic Planning Workshop on Global Ocean Issues in Marine Areas Beyond National Jurisdiction in the Context of Climate Change*, Nice, 2008, available at: <http://www.globaloceans.org/globalconferences/2008/pdf/High-Seas-PB-April9.pdf> accessed 11 September 2017. See also M. Balgos et al., *Executive Summary on the Workshop on Governance of Marine Areas Beyond National Jurisdiction: Management Issues and Policy Options*, Singapore, 2009.

Principle 1: Conditional Freedom of the Seas

Article 87 of UNCLOS recognizes six 'freedoms' of the high seas, namely:

a) Freedom of Navigation
b) Freedom of Overflight
c) Freedom to Lay Submarine Cables/Pipelines
d) Freedom to Construct Artificial Islands/Installations
e) Freedom of Fishing
f) Freedom of Scientific Research

What is often forgotten however is that these are not absolute rights but are subject to a number of limitations and corresponding duties upon which their legal exercise is pre-conditioned. Unfortunately these duties and conditions tend to be forgotten. An object lesson perhaps is freedom of fishing. Under Article 116 of UNCLOS all States have the right for their nationals to engage in fishing on the high seas, subject to three conditions:

(a) their treaty obligations
(b) the rights and duties, … [and] interests of coastal States …
(c) the provisions of this section.

So this is by no means an unfettered and absolute right. It is subject to all the treaty obligations that the flag state may have contracted by its membership of global and regional treaty regimes including regional and species fisheries conservation and management treaties. It is subject to the whole slew of rights and duties that it may owe to, or be due as, a coastal State (b) and finally the provisions of Articles 116–120 (i.e. section 2 of Part VII of UNCLOS) that, briefly summarized, include obligations to take measures for their own nationals for the conservation of the living resources of the high seas (Article 117); to co-operate with other States in conservation and management of those resources (Article 118); and to base those measures on the best scientific evidence available, environmental and economic factors and 'generally recommended international minimum standards' (Article 119). So, although UNCLOS talks of freedom of fishing, it is worth remembering that this is a conditional freedom. Similar conditions condition the exercise of the other freedoms and one can, and should, therefore talk about conditional high seas freedoms, rather than absolute rights.

Principle 2: Protection and Preservation of the Marine Environment

In relation to the marine environment, UNCLOS introduced in its Article 192 a major new principle: an unprecedented, unqualified and potentially onerous obligation on all states to 'protect and preserve the marine environment'. It also contains more specific obligations to protect and preserve rare or fragile species and ecosystems in all parts of the marine environment, as well as the habitat of depleted, threatened or endangered species and other forms of marine life.[12] Article 192, however, is a general obligation that extends further than simply the avoidance of deliberate and/or obvious damage, so as to include active measures to maintain or improve the present condition of the marine environment,[13] as well as to co-operate to this end.[14] So, the general obligations of Article 192 *et al.* reflect both the responsibility to conserve marine ecosystems as well as to prevent marine pollution.

Principle 3: International Cooperation

The principle that members of the international community have a duty of international co-operation is well established in general international law. In the famous 1970 UNGA Declaration of Principles of International Law that is generally accepted as being declaratory of customary international law, the General Assembly declared that:

> All States have the duty to cooperate with one another [...] in the various spheres of international relations, in order to maintain international peace and security and to promote international economic stability and progress [...][15]

12 Art. 194(5) UNCLOS.

13 M. Nordquist et al. (eds.), *The 1982 Law of the Sea Convention: A Commentary*, vol. IV, 1990, p. 40.

14 Art. 197 UNCLOS. See principle 3 below.

15 UNGA Res. 2625 (XXV) (24 October 1970). Adopted without a vote.

Various international instruments governing environment and natural resource conservation also include an obligation to co-operate. An example already referred to above in relation to the high seas is Article 117, which provides that 'All States have the duty to take or co-operate with other States in taking, such measures for their respective nationals as may be necessary for the conservation of the living resources of the high seas'. Of much wider application is Principle 7 of the 1992 Rio Declaration adopted at the UN Conference on Environment and Development (UNCED), which obliges States to 'co-operate in a spirit of global partnership to conserve protect and restore the health and integrity of the Earth's ecosystem'. And of course Rio Principle 27, which requires that 'States and people shall cooperate in good faith and in a spirit of partnership in the fulfillment of the principles embodied in this Declaration and in the further development of international law in the field of sustainable development'.

Principle 4: Science-Based Approach to Management

As we have seen above, UNCLOS mandates a science-based approach to management in that its Article 119 requires States to base their fisheries conservation and management measures on 'the best scientific evidence available' as well as environmental and economic factors and 'generally recommended international minimum standards'. These same obligations are reflected in the 1995 UN Fish Stocks Agreement (UNFSA) which requires States to ensure that in adopting measures to ensure the long term sustainability of straddling and highly migratory fish stocks, coastal States and States fishing on the high seas shall 'ensure that such measures are based on the best scientific evidence available and designed to maintain or restore stocks at levels capable of maximum sustainable yield'.[16] In fact the precautionary methodology in the UNFSA that is set out in Article 6 and Schedule II and discussed below, requires that scientific reference points are established for target species 'derived from an agreed scientific procedure' to constrain harvesting within safe biological limits. Many contemporary fisheries and natural resource management agreements, such as CCAMLR and the NEAFC,[17] have incorporated these approaches.

16 Art. 5 UNFSA.

17 Convention on Future Multilateral Cooperation in North-East Atlantic Fisheries (18 November 1980, entered into force 17 March 1982) 1285 UNTS 129.

Principle 5: The Precautionary Approach

The widespread acceptance of the precautionary principle or approach is one of the most important and distinctive developments in the international law relating to the environment and the management of natural resources.[18] In November 1990, the UN Secretary General expressly recognized the 'considerable significance' of the precautionary principle for future approaches to marine environmental protection and resource conservation.[19] Since then it has featured in virtually all international environmental treaties and policy declarations, most notably those relating to the marine environment and resources.[20] Agenda 21 mandates 'new approaches to marine and coastal area management [...] that are integrated in content and precautionary and anticipatory in ambit'.[21] Principle 15 of the UNCED Rio Declaration provides that:

> In order to protect the environment, the precautionary approach shall be widely applied by States according to their capabilities. Where there are threats of serious of irreversible damage lack of full scientific certainty

18 And a particular interest of mine, see eg: D. Freestone, 'The Precautionary Principle' in R. Churchill and D. Freestone (eds.), *International Law and Global Climate Change*, Graham & Trotman, 1991, pp. 21–40; D. Freestone and E. Hey, *The Precautionary Principle in International Law: the Challenge of Implementation*, Kluwer Law International, 1996; D. Freestone, 'Implementing Precaution Cautiously: The Precautionary Approach in the 1995 Straddling Stocks Agreement' in E. Hey (ed.), *Developments in International Fisheries Law*, Kluwer Law International, 1999, pp. 287–325; D. Freestone, 'Caution or Precaution: "A rose by any other name ...?"', *Yearbook of International Environmental Law*, vol. 12, 1999, pp. 25–32.

19 'Report of the UN Sec Gen. on the Law of the Sea' (19 November 1990), p. 20, para. 60.

20 Examples of such post Rio documents include: 1993 Ministerial Declaration on the Protection of the Black Sea see *Environmental Policy and Law*, vol. 23, 1993, pp. 235–236; UN Framework Convention on Climate Change, 31 ILM 848 (1992); Convention on Biological Diversity, 31 ILM 818 (1992); 1992 Helsinki Convention on the Protection of the Baltic Sea Area 3 YBIEL (1992) and in 8 IJMCL 215 (1993), with commentary by P. Ehlers at 191; 1992 Helsinki Convention on the Protection and Use of Transboundary Watercourses and Lakes, 31 ILM 1312 (1992); the 1992 Maastricht Treaty on European Union 31 ILM 247 (1992); 32 ILM 1693 (1993); the 1992 Paris Convention on the Protection of the Marine Environment of the North-East Atlantic, 8 IJMCL, pp. 50–76 (1993), and of course the 1995 UN Fish Stocks Agreement.

21 Agenda 21, para. 17.1.

> shall not be used as a reason for postponing cost-effective measures to prevent environmental degradation.[22]

Precaution is a key strategy of the UNFSA (an implementing agreement to UNCLOS). Article 6 requires that, to protect marine living resources as well as preserve the marine environment, the precautionary approach shall be applied widely to conservation, management and exploitation measures. It requires caution when information is uncertain, unreliable or inadequate; moreover, the absence of adequate scientific information shall not be used as a reason for postponing or failing to take conservation and management measures. Annex II of the Agreement sets out guidelines for the application of this approach in relation to the conservation and management of relevant fish stocks. This was the first time that an operational precautionary methodology for fisheries management had been set out in a treaty, but it has since been widely incorporated into the practice of many natural resource management regimes. Precaution is also a key component of the practice of the 1972 London Convention and is expressly included in its 1996 Protocol.

In a landmark Advisory Opinion in 2011,[23] the Seabed Disputes Chamber of the International Tribunal for the Law of the Sea (ITLOS) identified what it referred to as the 'legal obligation' to apply the precautionary approach as found in Principle 15 of the Rio Declaration.[24] Precaution is required by the International Seabed Authority nodules and sulfides regulations, but the Chamber went further, seeing the approach as 'an integral part of the general obligation of due diligence of sponsoring States, which is applicable even outside the scope of the Regulations' by requiring actions where scientific evidence on the scope and harmful impact of the activities is insufficient but 'there are plausible indications of potential risks'.[25] Perhaps most significant, the Chamber recognized 'a trend towards making this approach part

---

22 For the documents adopted at UNCED see UN Doc. A/Conf.151/26 (vols. 1-V) (12 August 1992).

23 *Responsibilities and Obligations of States Sponsoring Persons and Entities with Respect to Activities in the Area*, Advisory Opinion, ITLOS Reports 2011, p. 10 [A.O.].

24 Note that the equivocal wording of Principle 15 requires states only to introduce 'cost-effective measures' 'according to their capabilities.' Rio Declaration on Environment and Development, UN Doc. A/CONF.151/5/Rev.1 (13 June 1992) [Rio Declaration], reprinted in 31 ILM 874 (1992). This wording was reportedly introduced by the United States at the 1992 Rio Conference.

25 A.O., para. 131.

of customary international law',[26] as it found in the *Pulp Mills* judgment and which its present opinion further supports.[27]

Principle 6: The Ecosystem Approach

Perhaps the first legal instrument to espouse an ecosystem approach is the 1980 Convention for the Conservation of Antarctic Marine Living Resources (CCAMLR). As Kiss and Shelton point out 'it considers the interrelationship between all species and their particular physical environment' and its coverage is 'uniquely based on a biological boundary', namely waters south of the Atlantic convergence.[28] The ecosystem approach to natural resource management begins to be reflected more widely in legal and policy instruments after the adoption by the UNGA of the World Charter for Nature in 1982.[29] This called on states to protect representative ecosystems but also mandated that ecosystems and species exploited by mankind should be managed so as not to endanger co-existing ecosystems and species. By 1992 it is possible to see this approach reflected in both Agenda 21 and the Convention on Biological Diversity.

From a marine perspective it is most obviously incorporated in the UNFSA, which together with the precautionary approach also requires that its parties assess the impacts of fishing, other human activities and environmental factors on target stocks and species belonging to the same eco-system or associated with or dependent upon the target stocks. Pursuant to Article 5(d) and (e), once such assessment has taken place member states shall 'adopt, where necessary, conservation and management measures for species belonging to the same ecosystem or associated with or dependent upon the target stocks, with a view to maintaining or restoring populations of such species above levels

---

26 A.O., para. 135.

27 See, e.g., D. Freestone & E. Hey, *The Precautionary Principle And International Law: The Challenge Of Implementation* Kluwer Law International, 1996; D. Freestone, 'The Marine Environment', in J.B. Wiener et al. (eds.), *The Reality Of Precaution: Comparing Risk Regulation In The United States And Europe*, RFF Press, 2011, p. 177.

28 A. Kiss and D. Shelton, *International Environmental Law*, Transnational Publishers, 2004, p. 645.

29 UNGA Res. 37/7 (28 October 1982), supported by 111 States (18 abstaining). See also W.E. Burhenne and W.A. Irwin, *The World Charter for Nature: a Background Paper*, E. Schmidt, 1983; International Council of Environmental Law, *Commentary on the World Charter for Nature*, IUCN Environmental Law Centre, Bonn, 1986.

at which their reproduction may become seriously threatened'. A number of regional and species fisheries agreements now reflect this approach.[30]

At the 2001 Reykjavik Conference on Responsible Fisheries in the Marine Ecosystem, organized by FAO and the Government of Iceland, states recognized in the final Declaration that sustainable fisheries management incorporating ecosystem considerations entails taking into account the impacts of fisheries on the marine ecosystem and the impacts of the marine ecosystem on fisheries. They also recognized the clear need to introduce immediately effective management plans with incentives that encourage responsible fisheries and sustainable use of marine ecosystems, including mechanisms for reducing excessive fishing efforts to sustainable levels and declared that the prevention of adverse effects of non-fisheries activities on the marine ecosystems and fisheries requires action by the relevant authorities and other stakeholders.[31]

Principle 7: Sustainable and Equitable Use

Just as the UNFSA illustrates the way that international law has responded to the challenges of the modern age by recognizing new principles and concepts such as precaution, other international legal instruments now increasingly recognize the new paradigm of "sustainable use" or "sustainable development."

First brought into the international arena by the 1987 Brundtland Commission on Environment and Development,[32] sustainable development has been hailed as a basic paradigm for the twenty-first Century. Sustainable development, as defined by the Brundtland Commission is 'development that meets the needs of the present without compromising the ability of future generations

30 The South-East Atlantic Fisheries Organization (SEAFO) was established by the convention signed in Windhoek, Namibia, 20 April 2001 (2002) 41 ILM 257: it is the first general fisheries convention to have been negotiated after the 1995 UNFSA and to reflect its requirements. Similarly, the Commission for the Conservation and Management of Highly Migratory Fish Stocks in the Western and Central Pacific Ocean (WCPFC) was established by the Convention and opened for signature at Honolulu , 5 September 2000, (2001) 40 ILM 277. The Convention was the first regional tuna fisheries agreement to be adopted after the conclusion of the 1995 UNFSA, and it also reflects its requirements.

31 The Reykjavik Conference on Responsible Fisheries in the Marine Ecosystem was held in Reykjavik, Iceland, from 1 to 4 October 2001. The Conference adopted the *Reykjavik Declaration on Responsible Fisheries in the Marine Ecosystem*, available at <http://www.fao.org/docrep/005/Y2198T/y2198t01.htm> accessed 11 September 2017.

32 'Report of the World Commission on Environment and Development: Our Common Future', annexed to UNGA Res. 42/427 (4 August 1987).

to meet their own needs'. It thus reinforces the equitable notion of fairness or equity in relation to the needs of present and future generations as balanced by environmental limits and goals. The principle was included as Principle 4 of the Rio Declaration, and permeates other principles, Agenda 21 and numerous other instruments. In 1997 it was considered by the International Court of Justice in the *Gabčíkovo-Nagymaros Case* between Hungary and Slovakia. Although the majority of the Court did not endorse the famous separate opinion of Judge Christopher Weeramantry that sustainable development was a principle of customary international law, it did however recognize the 'need to reconcile economic development with protection of the environment [...] aptly expressed in the concept of sustainable development'.[33] A commitment to sustainable use can now be found in a raft of international instruments, including those relating to ocean use such as the UNFSA,[34] the 1995 FAO Code of Conduct for Responsible Fisheries and the 2001 Reykjavik Declaration on Responsible Fisheries in the Marine Ecosystem.[35]

Sustainable use of fisheries is also included in the commitments of the world community in the 2002 Johannesburg World Summit on Sustainable Development Plan of Implementation. A well-publicized aspect of this is the disproportionate overcapitalization and use of state subsidies in the fisheries sector which decreases the ability of developing countries as new entrants to benefit from fisheries (intra-generational equity) and diminishes future options for sustainable fisheries (inter-generational equity).[36]

Principle 8: Public Availability of Information

Principle 10 of the Rio Declaration recognizes that 'Environmental issues are best handled with the participation of all concerned citizens, at the relevant level [...] States shall facilitate and encourage public awareness and participation by making information widely available'. These hortatory provisions have

33 *Gabčíkovo-Nagymaros Project* (Hungary v. Slovakia), Judgment, ICJ Reports 1997, p. 7, p. 78 at para. 140.

34 Art. 5(a) UNFSA. States that are party to the Agreement are for example obliged to '(a) Adopt conservation and management measures to ensure long-term sustainability and promote the objective of their optimum utilization'.

35 In relation to fisheries the UN Fish Stocks Agreement, the FAO Code of Conduct on Responsible Fisheries, as well as the 2001 Reykjavik Declaration on Responsible Fisheries in the Marine Ecosystem, available at <ftp://ftp.fao.org/fi/DOCUMENT/reykjavik/y2198too_dec.pdf> accessed 11 September 2017.

36 FAO/World Bank, *The Sunken Billions*, 2008.

been given important legal substance by that 1998 ECE Aarhus Convention on Access to Information, Public Participation in Decision Making in Environmental Matters (Aarhus).[37] While Aarhus is a convention concluded under the auspices of the UN Economic Commission for Europe (ECE), it is highly unusual in that it is open for accession by any other UN Member state, even if not an ECE member, with approval of the Meeting of the Parties.[38] While Aarhus is directly relevant to national environmental decision-making, it also represents the "gold standard" for the implementation of the aspirations of Principle 21 of the Rio Declaration and many of the European nations that are party to other international agreements concerning the high seas – such as the 1972 London Convention and its 1996 Protocol as well as Regional Fishery Management Organizations (RFMOs) – are also party to Aarhus. This principle leads directly to the following one.

Principle 9: Transparent and Open Decision Making Processes

Transparency and openness in the conduct of the work of international and intergovernmental processes is now becoming the norm. Treaty-based organizations such as the International Maritime Organization (IMO) and the meetings of the Conference of the Parties to multilateral conventions such as the Convention on Biological Diversity, while acknowledging that States are the primary players, do accord access to other non-state parties. Background papers and secretariat papers are commonly distributed to state and non-state participants. Despite the fact that the biological resources of the high seas could be regarded as a common good, this is not the case – or certainly has not been the case in the past – for high seas fisheries management bodies. It is Article 12 of UNFSA which for the first time introduces an obligation on its States parties to provide for 'transparency in the decision-making process and other activities or sub-regional and regional fisheries management organizations and arrangements'. Article 12(2), which has already been adopted by some fisheries bodies and may be regarded as best international practice, specifically provides that:

37 Signed 25 June 1998, entered into force 30 October 2001. It has 41 State parties and the EC. The EC has already started to reflect Aarhus principles in Community, notably Directive 2000/60/EC, Water Framework Directive.

38 Under the provisions of Art 19(3).

> Representatives from other international organizations and representatives from non-governmental organizations concerned with straddling fish stocks and highly migratory fish stocks shall be afforded the opportunity to take part in meetings of subregional and regional fisheries management organizations and arrangements as observers or otherwise, as appropriate, in accordance with the procedures of the organization or arrangement concerned. Such procedures shall not be unduly restrictive in this respect. Such intergovernmental organizations and non-governmental organizations shall have timely access to the records of such organizations and arrangements, subject to the procedural rules on access to them.

Principle 10: Responsibility of States as Stewards of the Global Marine Environment

Principle 21 of the 1972 Stockholm Declaration provides that:

> States, have in accordance with the Charter of the United Nations and the principles of international law, the sovereign right to exploit their own resources pursuant to their own environmental policies, and the responsibility to ensure that activities within their jurisdiction or control do not cause damage to the environment of other states or of areas beyond the limits of national jurisdiction.

These rights and obligations are repeated virtually verbatim in Rio Principle 2. For our purposes a simpler statement of a principle, derived directly from these words and applicable to the high seas and which would be widely regarded as a principle of customary international law, would read as follows:

> States […] have the responsibility to ensure that activities within their jurisdiction and control do not cause damage to the environment […] of areas beyond the limits of national jurisdiction.

This concept of responsibility reflects a proactive obligation that would support a number of proposals that have been put forward for a form of stewardship role in protecting the resources of areas beyond national jurisdiction.[39]

39 P.H. Sand, 'Sovereignty Bounded: Public Trusteeship for Common Pool Resources?', *Global Environmental Politics*, vol. 4, 2004, pp. 47–71; R. Rayfuse and R. Warner, 'Securing a

It was that same concept of responsibility that the drafters seem to be trying to capture in the 1995 FAO Code of Conduct for Responsible Fisheries. That, and similar provisions of the UNFSA and the 1993 FAO Compliance Agreement, require flag States to supervise properly the activities of their fishing vessels when on the high seas. And yet the continued major threats of illegal, unregulated and unreported (IUU) fishing demonstrate that flag States are simply not exercising this sort of control. Responsibility in this sense involves an obligation on States not merely to regulate vessels flying their flag operating on the high seas – and many States seem unable to manage that – but also its nationals, captains, crews, owners and investors – all in the value chain of activities which do, or might, cause harm to the environment in areas beyond national jurisdiction. The FAO has developed principles for audits of Flag State Responsibility.

It is paradoxical that in looking for the articulation of this wider concept of responsibility in an existing treaty, the best example perhaps is to be found in Article VI of the 1979 Treaty on Principles Governing the Activities of States in the Exploration and Use of Outer Space, including the Moon and Celestial Bodies. It reads:

> States Parties to the Treaty shall bear international responsibility for national activities in outer space, including the Moon and other celestial bodies, whether such activities are carried on by governmental agencies or by non-governmental entities, and for assuring that national activities are carried out in conformity with the provisions set forth in the present Treaty. The activities of non-governmental entities in outer space, including the Moon and other celestial bodies, shall require authorization and continuing supervision by the appropriate State Party to the Treaty.

Writing in 2008 I bemoaned the paradox that it would be tragic if we were left with the impression that the international community is more concerned with the Moon than with our own Blue Planet.[40] However, since that date, the ITLOS has begun to develop a very important line of cases developing a similar concept.

Sustainable Future for the Oceans beyond National Jurisdiction', *International Journal of Marine and Coastal Law*, vol. 23, 2008, pp. 399–422.

40 See D. Freestone, 'Modern Principles of High Seas Governance: The Legal Underpinnings', *International Environmental Policy and Law*, vol. 39 no. 1, 2009, p. 49. The Moon Treaty has not commanded a large number of parties: 13 States have ratified and a further four had signed only.

On 1 February 2011, the Seabed Disputes Chamber (Chamber) of the ITLOS unanimously adopted a historic Advisory Opinion, *Responsibilities and Obligations of States Sponsoring Persons and Entities with Respect to Activities in the Area*.[41] Among other things, the Chamber defines the responsibilities and obligations of States that sponsor activities in the seabed area beyond national jurisdiction (the Area) and the extent of the sponsoring State's liability for the failure by an entity it has sponsored to comply with UNCLOS. The request for an Advisory Opinion arose in the context of guidance sought by two developing States – Nauru and Tonga – in respect of their obligations regarding possible sponsorship of deep seabed exploration and mining activities.

On the primary question of legal responsibilities and obligations of a State that sponsors seabed activities, the text of UNCLOS is relatively clear. Article 139(1) reads:

> States Parties shall have the responsibility to ensure that activities in the Area, whether carried out by States Parties, or state enterprises or natural or juridical persons which possess the nationality of States Parties or are effectively controlled by them or their nationals, shall be carried out in conformity with this Part.

States Parties are obliged to assist the Authority in this respect.[42] Moreover, under Article 4(4), Annex III of UNCLOS, State sponsors shall also, 'pursuant to Article 139, have the responsibility to ensure, within their legal systems, that a contractor so sponsored shall carry out activities in the Area in conformity with the terms of its contract and its obligations under this Convention'. Thus, ruled the Chamber, the sponsoring State must take all measures necessary to ensure the contractor's compliance and those measures must be incorporated in that State's legal system.[43]

Having recognized that Article 139 contains an "obligation to ensure," the Chamber itemized the constituent elements of this obligation, pointing out that it is an obligation of conduct rather than result. More specifically, the obligation does not require the contractor's compliance in every case.[44] It

41 See A.O., n. 23. For comment, D. Freestone, 'Responsibilities and Obligations of States Sponsoring Persons and Entities with respect to Activities in the Area', *American Journal of International Law*, vol. 105, 2011, pp. 755–761.

42 UNCLOS Art. 153(4).

43 A.O., para. 118.

44 Ibid., paras. 109–10.

is analogous to the obligation of due diligence and of conduct that the International Court of Justice had previously found in the 2010 *Pulp Mills* case.[45]

Other obligations relevant to meeting the due diligence obligation to ensure (termed "direct obligations" by the Chamber) include "best environmental practices", which are required by the ISA sulfides regulations and the standard clauses for exploration contracts.[46] Sponsoring States are also to provide technical and financial guarantees regarding compliance by the contractor with emergency orders for protection of the marine environment, as well as to make available financial recourse for prompt and effective compensation in the event of damage caused by marine pollution.[47] The additional obligation to conduct environmental impact assessments was found by the Chamber to extend beyond the scope of the ISA regulations.[48]

On the wider and controversial question of the treatment of developing States, the Chamber unequivocally endorsed the principle of equality. It recognized that enterprises in developed States might establish companies in developing states and then obtain their nationality to avoid more stringent regulations, and that the spread of such sponsoring "States of convenience" (similar to flags of convenience for ships) would jeopardize the application of the highest standards of protection.[49]

In the second important case on 2 April 2015, the ITLOS delivered its first ever full Advisory Opinion in response to four questions submitted by the State parties to the West Africa Sub-Regional Fisheries Commission (SRFC).[50] The Opinion clarifies a number of important issues regarding the obligations of flag States to control the activities of their vessels conducting fishing activities in the EEZs of other States and their potential liability for failure to do this effectively. It addresses the obligations of an international organization, such as the EU, where vessels flying the flags of the Member States of the organization

45 *Pulp Mills on the River Uruguay* (Argentina v. Uruguay), Judgment, ICJ Reports 2010, para. 187.

46 ISA, Standard Clauses for Exploration Contract, available at <http://www.isa.org.jm/en/documents/mcode> accessed 11 September 2017.

47 A.O., paras. 122–23, 136, 138, 140.

48 A.O., paras. 145, 148.

49 A.O., para. 159.

50 *Request for an Advisory Opinion submitted by the Sub-Regional Fisheries Commission (SRFC)*, Advisory Opinion, ITLOS Reports 2015, available at <https://www.itlos.org/cases/list-of-cases/case-no-21/> accessed 11 September 2017. For comment see D. Freestone, 'Case 21: Request for an Advisory Opinion submitted by the Sub-Regional Fisheries Commission (SRFC), International Tribunal for the Law of the Sea', *Asia-Pacific Journal of Ocean Law and Policy*, vol. 1, 2016, pp. 126–133.

are fishing under the terms of an access agreement between the organization and a coastal State or States. It also considers the obligations of cooperation under UNCLOS between coastal States and others engaged in fishing for highly migratory species in the region. The Opinion represents an important further application of the requirements of "due diligence" which had been developed by the ITLOS Seabed Disputes Chamber in its 2011 Advisory Opinion and discussed above.

There has long been debate as to the relevance of the duties of the Flag State set out in Article 94 of UNCLOS to vessels engaged in fishing. Article 94(1) appears to concern itself primarily with the internal order of vessels flying a flag. It provides that 'Every State shall effectively exercise its jurisdiction and control in administrative, technical and social matters over vessels flying its flag' and goes on in 94(2): 'In particular every State shall: (a) maintain a register [...]; and (b) assume jurisdiction [...] in respect of administrative, technical and social matters concerning the ship'. The remainder of the article deals with construction and certification of seaworthiness, qualifications of masters, crew etc. – important issues, but not specifically related, it has been argued, to the way in which the vessel conducts the activities in which is involved (fishing, marine research etc.).

The interpretation of Article 94 of UNCLOS by the ITLOS ends this controversy; for it decided that the use of the words 'In particular' in Article 94(2) indicated that the list of flag State responsibilities that followed was not exhaustive but only indicative. Hence, it said 'as far as fishing activities are concerned, the flag State, in fulfilment of its responsibility to exercise effective jurisdiction and control in administrative matters, must adopt the necessary administrative measures to ensure that fishing vessels flying its flag are not involved in activities which will undermine the flag State's responsibilities under the Convention in respect of the conservation and management of marine living resources'.[51] Moreover the Tribunal reminded us that it had already found in the *Southern Bluefin Tuna Cases*,[52] that a flag State's obligation under Article 192 of UNCLOS to 'protect and preserve the marine environment' include 'conservation of the living resources of the sea'. Therefore flag States are obliged to take the necessary measures to ensure that their nationals and vessels flying their flag are not involved in IUU fishing activities in the EEZ of another State.[53]

---

51 A.O., para. 119.

52 *Southern Bluefin Tuna Cases* (New Zealand v. Japan; Australia v. Japan), Provisional Measures, Order of 27 August 1999, ITLOS Reports 1999, p. 295, para. 70.

53 A.O., para. 124.

When considering exactly what such "necessary measures" might be in relation to flag States overseeing their fishing boats, the ITLOS drew attention to the standard of due diligence required of States in the 2011 Advisory Opinion of the Seabed Disputes Chamber. Although the relationship between a State and an entity that it is sponsoring for mineral exploration or exploitation activities in the Area is not entirely comparable with that between a flag State and its fishing vessels, nevertheless it felt that the same principles are applicable.

Hopefully these cases are the beginning of an important and developing jurisprudence, in which the international judiciary will take even further the concept of the basic responsibilities of States regarding the marine environment.

Bibliography

Balgos, M. et al., *Executive Summary on the Workshop on Governance of Marine Areas Beyond National Jurisdiction: Management Issues and Policy Options*, Singapore, 2009.

Burhenne, W.E., Irwin, W.A., *The World Charter for Nature: a Background Paper*, E. Schmidt, 1983.

Cicin-Sain, B., Freestone, D., *Report from the Strategic Planning Workshop on Global Ocean Issues in Marine Areas Beyond National Jurisdiction in the Context of Climate Change*, Nice, 2008.

Freestone, D., 'Case 21: Request for an Advisory Opinion submitted by the Sub-Regional Fisheries Commission (SRFC), International Tribunal for the Law of the Sea', in *Asia-Pacific Journal of Ocean Law and Policy*, vol. 1, no. 1, 2016, p. 126.

Freestone, D., 'Caution or Precaution: "A rose by any other name ...?"', *Yearbook of International Environmental Law*, vol. 12, 1999, p. 25.

Freestone, D., 'Governance of Areas beyond National Jurisdiction: An Unfinished Agenda?', in *Law of the Sea: UNCLOS as a Living Treaty*, British Institute of International and Comparative Law, 2016, p. 231.

Freestone, D., 'Implementing Precaution Cautiously: The Precautionary Approach in the 1995 Straddling Stocks Agreement' in *Developments in International Fisheries Law*, Kluwer Law International, 1999, p. 287.

Freestone, D., 'Modern Principles of High Seas Governance: The Legal Underpinnings', in *International Environmental Policy and Law*, vol. 39, no. 1, 2009, p. 44.

Freestone, D., 'Responsibilities and Obligations of States Sponsoring Persons and Entities with respect to Activities in the Area', in *American Journal of International Law*, vol. 105, 2011, p. 755.

Freestone, D., 'The Marine Environment', in *The Reality Of Precaution: Comparing Risk Regulation In The United States And Europe*, RFF Press, 2011, p. 177.

Freestone, D., 'The Precautionary Principle' in *International Law and Global Climate Change*, Graham & Trotman, 1991, p. 21.

Freestone, D., Hey, E., *The Precautionary Principle in International Law: The Challenge of Implementation*, Kluwer Law International, 1996.

Kiss, A., Shelton, D., *International Environmental Law*, Transnational Publishers, 2004.

Nordquist, M. et al., *The 1982 Law of the Sea Convention: A Commentary*, vol. 4, 1990, p. 40.

Rayfuse, R. et al., *Co-Chairs' Report of Workshop on High Seas Governance for the 21st Century*, New York City, 2007.

Rayfuse, R., Lawrence, M. and Gjerde, K., 'Ocean Fertilization and Climate Change: the Need to Regulate Emerging High Seas Uses', in *International Journal of Marine and Coastal Law*, vol. 23, no. 2, 2008, p. 297.

Rayfuse, R., Warner, R., 'Securing a Sustainable Future for the Oceans beyond National Jurisdiction', in *International Journal of Marine and Coastal Law*, vol. 23, no. 3, 2008, p. 399.

Sand, P.H., 'Sovereignty Bounded: Public Trusteeship for Common Pool Resources?', in *Global Environmental Politics*, vol. 4, no. 1, 2004, p. 47.

Warner, R., Gjerde, K. and Freestone, D., 'Regional Governance for Fisheries and Biodiversity', in *Governance for Fisheries and Marine Conservation: Interactions and co-evolution*, Wiley-Blackwell, 2014, p. 211.

PART 2

Sustainable Ocean Resource Governance – The Specific Fields of Application

CHAPTER 8

The Area: Common Heritage of Mankind, Sponsoring States of Convenience and Developing States

Edwin Egede

1 Introduction

Based on writings of scholars, notably John Mero,[1] and the inspiring speech at the United Nations General Assembly by the then Maltese Ambassador to the United Nations, Arvid Pardo,[2] painting a picture of huge resources in the deep seabed beyond national jurisdiction (the Area) available to be easily picked up, the third United Nations Conference on the Law of the Sea (UNCLOS III) adopted a relatively comprehensive but complex regime dealing with the Area, which was incorporated into Part XI of UNCLOS 1982.[3] As a result of the objections by certain developed States to aspects of Part XI of the UNCLOS, the Agreement relating to the implementation of Part XI of the United Nations Convention on the Law of the Sea of 10 December 1982 (the 1994 Agreement) was negotiated and adopted in 1994. The latter Agreement, which varied some of the provisions of Part XI, was in essence an amending instrument and it paved the way for all the objecting developed States, except the USA, to eventually become Parties to the UNCLOS and the 1994 Agreement.[4] With the entry into force of the UNCLOS and the 1994 Agreement on the 16 November

* The author conveys his thanks to the Scientific Advisory Board of the Hamburg International Environmental Law Conference for the kind invitation to present this paper at the 3rd Hamburg International Environmental Law Conference in April 2016. He also expresses special thanks to Prince Emmanuel for his usual support and inspiration – you are indeed a friend who sticks closer than a brother does.

1 J.L. Mero, *The Mineral Resources of the Sea*, Elsevier Publishing, 1968.

2 UNGA GAOR, 22nd Session (1 November 1967), A/C.1/PV.1515.

3 See Articles 133 to 191 of United Nations Convention on the Law of the Sea (adopted 10 December 1982, entered into force 16 November 1994) 21 ILM 1261 (1982) (UNCLOS).

4 See generally E. D. Brown, 'The 1994 Agreement on the Implementation of Part XI of the UN Convention on the Law of the Sea: Breakthrough to Universality?', *Marine Policy*, 1995, pp. 5–20. There are currently 167 parties to UNCLOS 1982 and 147 parties to the 1994 Implementation Agreement.

 | DOI 10.1163/9789004360273_009

1994 and 28 July 1996 respectively, the International Seabed Authority (ISA), an international organization established under Part XI of the UNCLOS, has been acting on behalf of the States Parties with regard to the Area and the resources therein, which are the Common Heritage of Mankind (CHM), to 'organize and control activities in the Area, particularly with a view to administering the resources of the Area.'[5] So far the ISA has put together a *Mining Code* comprising of: the Regulations on Prospecting and Exploration for Polymetallic Nodules in the Area;[6] the Regulations on Prospecting and Exploration for Polymetallic Sulphides in the Area;[7] the Regulations on Prospecting and Exploration for Cobalt-rich Ferromanganese Crusts in the Area;[8] and the Decision of the Assembly of the International Seabed Authority concerning overhead charges for the administration and supervision of exploration contracts,[9] along with certain recommendations for guidance of the contractors.[10] It is currently working on the Mining Code for Exploitation, which has been described as 'ultimate regulatory phase in developing the common heritage of mankind.'[11] Also, the ISA has entered into various contracts for exploration of polymetallic nodules, polymetallic sulphides and cobalt-rich ferromanganese crusts with

5 See Articles 136, 137 and 157 of UNCLOS.

6 International Seabed Authority (ISA), Regulations on Prospecting and Exploration for Polymetallic Nodules in the Area (adopted 13 July 2000, updated and adopted 25 July 2013).

7 ISA, Regulations on Prospecting and Exploration for Polymetallic Sulphides in the Area (adopted 7 May 2010).

8 ISA, Regulations on Prospecting and Exploration for Cobalt-rich Ferromanganese Crusts in the Area (adopted 27 July 2012).

9 ISA, Decision of the Assembly of the International Seabed Authority concerning overhead charges for the ad-ministration and supervision of exploration, ISBA/19/A/12 (22 October 2012).

10 See 'Recommendations for the Guidance of Contractors and Sponsoring States Relating to Training Programmes under Plans of Work for Exploration', ISBA/19/LTC/14 (12 July 2013); 'Recommendations for the Guidance of Contractors for the Assessment of the Possible Environmental Impacts Arising from Exploration for Marine Minerals in the Area', ISBA/19/LTC/8 (1 March 2013); 'Recommendations for the Guidance of Contractors for the Reporting of Actual and Direct Exploration Expenditure', ISBA/21/LTC/11 (14 April 2015); and 'Recommendations for the Guidance of Contractors on the Content, Format and Structure of Annual Reports', ISBA/21/LTC/15 (4 August 2015).

11 <https://www.isa.org.jm/legal-instruments/ongoing-development-regulations-exploitation-mineral-resources-area> accessed 1 July 2017.

a number of States, State enterprises and corporations, including developing States.[12]

One of the major concerns of the ISA, at the moment, is to ensure that the marine environment is protected from harmful effects that may occur during mining activities in the Area.[13] In 2011, the ITLOS Seabed Disputes Chambers, at the request of the Council of the ISA, gave its first advisory opinion regarding the responsibilities and obligations of States and their sponsored entities with respects to activities in the Area.[14] One of the issues raised by the chambers was the importance of safeguarding against so-called Sponsoring States of Convenience to protect the marine environment from unregulated or inadequately regulated Deep Seabed Mining (DSM) activities by sponsored contractors. Of course, the chambers expect that most, if not all, of these so-called Sponsoring States of Convenience would probably be from developing States. Therefore, this chapter explores some of the implications of the Advisory Opinion of the Chambers on developing States.

2 Developing States

The regime of the Area under Part XI of the UNCLOS 1982, as amended by the 1994 Implementation Agreement, is framed in such a way as to encourage

12 For the names of the various contractors see <https://www.isa.org.jm/deep-seabed-minerals-contractors?qt-contractors_tabs_alt=0#qt-contractors_tabs_alt> accessed 1 July 2017.

13 Article 145 of UNCLOS.

14 ITLOS Case No.17, Responsibilities and Obligations of States With Respect to Activities in the Area, Advisory Opinion, 1 February 2011, ITLOS Reports 2011, p. 10. For excellent analysis of this Advisory Opinion see D. French, 'From the Depths: Rich Pickings of Principles of Sustainable Development and General International Law on the Ocean Floor – the Seabed Disputes Chamber's 2011 Advisory Opinion', *International Journal of Marine and Coastal Law*, vol. 26, 2011, pp. 525–568; D. Freestone, 'Responsibilities and Obligations of State Sponsoring Persons and Entities With Respect to Activities in the Area', *American Journal of International Law*, vol. 105, 2011, pp. 755–760; D.K. Anton, R.A. Makgill and C.R. Payne, 'Seabed Mining – Advisory Opinion on Responsibility and Liability', *Environmental Policy and Law*, vol. 41, 2011, pp. 60–65; D.K. Anton, 'The Principle of Residual Liability in the Seabed Disputes Chamber of the International Tribunal for the Law of the Sea: The Advisory Opinion on Responsibility and Liability for International Seabed Mining (ITLOS Case No. 17)', *McGill International Journal of Sustainable Development Law & Policy*, vol. 7, 2011–2012, pp. 241–257; Y. Tanaka, 'Obligations and Liability of Sponsoring States Concerning Activities in the Area: Reflections on the ITLOS Advisory Opinion of 1 February 2011', *Netherlands International Law Review*, vol. 60, 2013, pp. 205–230.

developing States to also engage in DSM activities. For instance, it specifically requires that the effective participation of developing States in activities in the Area should be promoted.[15] It also provides a 'parallel' or site banking system of mining of the Area that requires a contractor in its application to the ISA for permission to carry out DSM activities to designate a tract in the Area sufficiently large and of appropriate estimated value to allow two mining operations, with the ISA having the first option to pick one of these sites which is reserved or banked (i.e. the reserved areas) for mining activities by the ISA in association with developing States.[16] Thus in different places the UNCLOS, as amended, refers to developing States with a view to make certain concessions to these States.[17] However, it fails to provide any clear guidance or criteria on how States are being classified as developing States. The interim report of the contractor that reviewed the activities under Article 154 had recommended that '[c]larification of the definition of 'developing State' as applied by the Authority would be helpful to satisfy Articles 148 and 150', and even suggested that the Council should seek an advisory opinion on the definition of what constitutes developing States for the purposes of the application of Part XI of the UNCLOS.[18] But, the Review Committee in its comments on the interim report indicated as follows:[19]

> A definition of 'developing State' for the purposes of applying the provisions of the Convention to such States would certainly be helpful. [However in] the opinion of the Review Committee, it seems doubtful whether the Article 154 Review Process would be the right forum for

---

15 Art. 148 of UNCLOS.

16 Art. 153 and Art. 8 of Annex III of UNCLOS and Section 1 para. 10 of the Annex to the Agreement relating to the Implementation of Part XI of the United Nations Convention on the Law of the Sea of 10 December 1982 (28 July 1994) (1994 Agreement). See E. Egede, *Africa and the Deep Seabed Regime: Politics and International Law of the Common Heritage of Mankind*, Springer, 2011, pp. 88, 190–199.

17 See for instance Art. 140(1), 143(3)(b), 144, 148, 150, 151(10), 152, 155(2), 161(1)(c, d) and (2)(b), 164(1), 173(2)(c) and section 1 para. 5(e), section 3, para. 15(c) and (d), section 4 para. 1(b) and (c), section 7 para. 1.

18 Seascape Consultants Ltd., *Periodic Review of the International Seabed Authority pursuant to UNCLOS Article 154. Interim Report*, 15 May 2016. pp. 24, 75, 81, 83, <https://www.isa.org.jm/sites/default/files/files/documents/isba-22a-crp.3-1.pdf> accessed 1 July 2017.

19 ISA, 'Periodic Review of the International Seabed Authority Pursuant to UNCLOS Article 154', Interim Report: Comments by the Review Committee, ISBA/22/A/CRP.3(2) (25 May 2016), para. 9, <https://www.isa.org.jm/sites/default/files/files/documents/isba-22a-crp3-2.pdf> accessed 1 July 2017.

> attempting to elaborate such a definition. Advice on this question might, however, be sought from the United Nations.

In essence, the Review Committee appeared to take the view that determining what exactly the definition of developing State is would fall within the ambit of the United Nations, the foremost high-level political intergovernmental forum. This would appear to indicate that their thinking was that the United Nations is the appropriate setting to engage with this issue, which seems to be more of a political issue rather than a legal one. For instance, we see the General Assembly adopting a resolution recognizing the graduation of countries like Equatorial Guinea and Vanuatu from the least developed country category.[20]

It is worth noting that the ITLOS Seabed Chamber in its unanimous advisory opinion did not endeavor to define what a developing State is, though it embarked on an extensive explanation of the meaning of terms such as responsibility, obligation, sponsorship, activities in the Area etc.[21] Could it be that the Chamber considered this to be outside its advisory opinion remit which is to 'give advisory opinions at the request of the Assembly or the Council on legal questions arising within the scope of their activities'?[22] Whilst, undoubtedly, a question seeking an advisory opinion on which States are developing States could certainly be framed into a legal type question by linking it to the relevant provisions of Part XI, it is really difficult to see how the Chamber's advisory opinion, by the very nebulous nature of the term 'developing States', could be based on law.[23] What legal instruments would the Chamber rely on to explain the meaning of the term 'developing State' in precise legal terms? Intrinsically, the term 'developing State' is not a legal concept, but rather a political one that falls more within the ambit of political economy. Thus, even though the phrase may be referred to in treaties, it would be an uphill task to seek to define such with precision, as can be seen by various attempts enunciated below to determine which precise States are developing States.

---

20 UNGA Res. 68/18 (9 December 2013).

21 See ITLOS Case No. 17, Responsibilities and Obligations of States With Respect to Activities in the Area, Advisory Opinion, 1 February 2011, ITLOS Reports 2011, p. 10.

22 Art. 191 of UNCLOS.

23 See *Accordance with International Law of the Unilateral Declaration of Independence in Respect of Kosovo*, Advisory Opinion (22 July 2010), ICJ Reports 2010, p. 403, para. 25; *Western Sahara*, Advisory Opinion (16 October 1975), ICJ Reports 1975, p. 12, para. 15, referred to by the Chambers.

There are no universally agreed criteria to determine which States are developing States.[24] For instance, there have been attempts to classify States as developing States based on their less developed industrial base and low human development index (HDI), as well as using the GNP per capita in comparison to other States. For example, the World Bank glossary states that developing countries (or States) are:

> […] countries with low or middle levels of GNP per capita as well as five high-income developing economies – Hong Kong (China), Israel, Kuwait, Singapore, and the United Arab Emirates. These five economies are classified as developing despite their high per capita income because of their economic structure or the official opinion of their governments. Several countries with transition economies are sometimes grouped with developing countries based on their low or middle levels of per capita income, and sometimes with developed countries based on their high industrialization. More than 80 percent of the world's population lives in the more than 100 developing countries.[25]

Furthermore, developing States in international politics may be said to be States that have membership of international political organizations, such as the G-77. This intergovernmental organization, actually made up of 134 States, consisting of a mixed bag of 'developing States' from different parts of the globe, describes itself as 'the largest intergovernmental organization of developing countries in the United Nations which provides the means for the countries of the South to articulate and promote their collective economic interests and enhance their joint negotiating capacity on all major international economic issues within the United Nations system, and promote South-South cooperation for development.'[26] It would appear that membership of such an organization, as the G-77, would confer on such States life-long membership of the developing States' 'club', even if over the years the fortunes of these States improve and they become high per capita income States. This appears to be a case of 'once a developing State, always a developing State.' In essence, there is an assortment of States that constitute developing States with diverse bed-fellows lumped together under this category. The danger of

---

24 See A. Sullivan and S.M. Sheffrin, *Economics: Principles in Action*, Prentice Hall 2003, p. 471.

25 <http://www.worldbank.org/depweb/english/beyond/global/glossary.html> accessed 1 July 2017.

26 <http://www.g77.org/doc/index.html> accessed 1 July 2017.

trying to get the international courts to seek to legalize an obviously political concept is that the courts would get embroiled in the politics of States. Thus an attempt to seek an advisory opinion of the Chambers of the ITLOS on the definition of developing States would be ill-advised and result in a legal quagmire.

What, however, is clear from the diversity of developing States is the difficulty in treating all developing States alike with such a wide variety of developing States, from the relatively 'developed' developing States to the 'least developed' developing States.[27] Thus it is difficult to have a one size fits all approach to engage with developing States and deep seabed mining. On the one hand, some so-called developing States are already engaged with deep seabed mining activities, either because they have the requisite technology and finance to do so (e.g. China, India and South Korea) or, in spite of not having the requisite technology and finance, have done so through sponsoring foreign Transnational Corporations (TNCs) that have incorporated local subsidiaries within their territory for this purpose (for instance, Island Pacific developing States, such as Nauru and Tonga).[28] On the other hand, there are some developing States (for instance those in Africa) that as yet are not in any way involved in deep seabed mining activities.[29] Despite this obvious difference in deep seabed mining capabilities, all these developing States technically enjoy exactly the same concessions as 'developing States' under Part XI of the UNCLOS 1982, as amended. However, it is important to note that despite the numerous concessions to developing States under Part XI, as amended, this does not apply to the responsibility regarding the protection of the marine environment. The ITLOS Seabed Disputes Chamber in its unanimous *Advisory Opinion on the Responsibilities and Obligations of States sponsoring persons and entities with respect to activities in the Area*, pointed out that:

---

27 Note that this variety of developing States is reflected in the UNCLOS. For instance, Art. 82(4) of UNCLOS referring to the distribution of the benefits and contributions made to the ISA from the outer CS refers to 'the interests and needs of developing States, particularly the least developed and the land-locked among them.' The complexity of fitting such diverse States into the category 'developing' has led the World Bank in its 2016 World Development Indicators to avoid using the term 'developing World'. See S. Sandhu, 'The World Bank has stopped using the term developing world', *i News*, May 18, 2016, <https://inews.co.uk/explainers/iq/no-longer-call-countries-developing/> accessed 1 July 2017.

28 <https://www.isa.org.jm/deep-seabed-minerals-contractors> accessed 1 July 2017.

29 See E. Egede, 'Institutional Gaps in the 2050 Africa's Integrated Maritime Strategy', *Journal of Ocean Law & Governance in Africa*, vol. 1, 2016, pp. 1–27, esp. pp. 19–26.

> [...] none of the general provisions of the Convention concerning the responsibilities (or the liability) of the sponsoring State 'specifically provides' for according preferential treatment to sponsoring States that are developing States. As observed above, there is no provision requiring the consideration of such interests and needs beyond what is specifically stated in Part XI. It may therefore be concluded that the general provisions concerning the responsibilities and liability of the sponsoring State apply equally to all sponsoring States, whether developing or developed.[30]

Thus, as regards the general obligation of protecting the marine environment, the Chamber stated that there could be no preferential treatment given to developing States. That being said, however, the Chamber appeared to soften this position by conceding that its observation did not exclude the possibility that rules setting out direct obligations of the sponsoring State could provide for different treatment for developed and developing sponsoring States. An example provided by the Chamber is the precautionary approach (or principle), which in its view, having been adopted in various treaties and international instruments that reflect its formulation in principle 15 of the Rio Declaration, had 'initiated a trend towards making this approach part of customary international law.'[31] Principle 15 of the Rio Declaration actually provides that the precautionary approach shall be applied by States 'according to their capabilities.'[32] In essence, as mentioned by the ITLOS Chamber, this means that 'the requirements for complying with the obligation to apply the precautionary approach may be stricter for the developed than for the developing sponsoring States.'[33] It also follows that apart from the differentiated responsibility between developed and developing States as regards the precautionary approach the expectation would be that there would also be a differentiated responsibility in this regard as between least developed developing States and more developed developing States, such as Brazil, China, India, Singapore and South Korea.[34] Albeit, the Chamber was quick to point out that the principle of differentiated responsibility does not apply to the obligation to follow 'best environmental practices' as required by the ISA exploration regulation.[35]

30 Para. 158; see also para. 156.

31 Paras 125, 135 and 161.

32 Para. 161.

33 Ibid.

34 P. Sands et al., *Principles of International Environmental Law*, 3rd Edition, Cambridge University Press, 2012, p. 236.

35 Ibid.

3 Sponsoring States of Convenience or Nauru-Tonga Model?

The UNCLOS provides that mining activities in the Area can be carried out either directly by States Parties or by entities sponsored by them. The Chamber points out that '[t]he notion of "sponsorship" is a key element in the system for the exploration and exploitation of the resources of the Area set out in the Convention.'[36] Article 139 dealing with the responsibility to ensure compliance and liability for any damages indicates that activities in the Area can either be carried out by 'States Parties, or State enterprises or natural or juridical persons which possess the nationality of States Parties or are effectively controlled by them or their nationals',[37] whereas Article 153 points out that activities in the Area may be carried out by 'States Parties, State enterprises or natural or juridical persons which possess the nationality of States Parties or are effectively controlled by them or their nationals, when sponsored by such States'.[38] In Annex III, the UNCLOS is emphatic that applicants, apart from the Enterprise,[39] shall only be qualified to apply to the ISA for approval of plans of work for activities in the Area if they have the nationality of or are effectively controlled by at least one of the States Parties.[40] In essence, this means that for an entity to be qualified to participate in mining activities in the Area it must either have the nationality of at least one State Party or must be effectively controlled by a State Party and in both cases it must be sponsored by such States. Traditionally, under international law, the determination of nationality falls within the domain of States' domestic jurisdiction. Oppenheim had affirmed this when he asserted: 'the conferment and deprivation of nationality is a right that International Law recognizes as being within the exclusive competence of States; but it is a right the abuse of which may be a ground for an international claim.'[41] Nationality issues in relation to individuals are relatively straightforward with complexities usually arising in cases of individuals with dual or multiple nationalities when there is contention amongst each of the national States as to which of them should be the preferred national State of such individual. The Nottebohm case declares that the prevailing view

36 Para. 74.

37 Article 139(1) of UNCLOS.

38 Article 153(2) of UNCLOS.

39 This is obviously subject to the modification in the 1994 Agreement which limits the ability of the Enterprise to engage in mining activities until certain conditions are fulfilled.

40 Article 4(1) of Annex III of UNCLOS.

41 L. Oppenheim, *International Law: A Treatise*, Longmans, 1937, p. 280. See *Nationality Decrees Issued in Tunis and Morocco*, Advisory Opinion (7 February 1923), PCIJ Ser. B No. 4.

is to prefer 'the real and effective nationality'[42] or the State with a 'genuine connection'[43] with the individual. However, it becomes a bit more complicated in the case of artificial juridical persons which have been given a distinct legal personality by law. The Barcelona Traction case, while identifying different ways by which States may confer nationality on these juridical entities, appeared to take the position that the established method in law is that such entities acquired the nationality of the State of incorporation and in whose territory the registered office is located.[44] In its decision in this case the ICJ stated:

> In allocating corporate entities to states for purposes of diplomatic protection, international law is based, but only to a limited extent, on an analogy with the rules governing the nationality of individuals. The traditional rule attributes the right of diplomatic protection of a corporate entity to the state under the laws of which it is incorporated and in whose territory it has its registered office. These two criteria have been confirmed by long practice and by numerous international instruments. This notwithstanding, further or different links are at times said to be required in order that a right of diplomatic protection should exist. Indeed, it has been the practice of some states to give a company incorporated under their law diplomatic protection solely when it had its seat (*siège social*) or management or center of control in their territory, or when a majority or a substantial proportion of the shares has been owned by nationals of the state concerned. Only then, it has been held, does there exist between the corporation and the state in question a genuine connection of the kind familiar from other branches of international law. However, in the particular field of the diplomatic protection of corporate entities, no absolute test of the 'genuine connection' has found general acceptance. Such tests as have been applied are of a relative nature, and sometimes links with one state have had to be weighed against those with another.[45]

The International Law Commission's (ILC) *Draft Articles on Diplomatic Protection* appears to accept the Barcelona Traction Case as generally consistent with customary international law and stated that: 'For the purposes of the diplomatic protection of a corporation, the State of Nationality means the State

42 *Nottebohm* (Liechtenstein v. Guatemala), Judgement Second Phase, ICJ Reports 1955, p. 22.

43 Ibid., p. 23.

44 *Barcelona Traction* (Belgium v. Spain), ICJ Reports 1970, p. 43.

45 Ibid., p. 42.

under whose law the corporation was incorporated.'[46] Some States are clear in case of DSM activities that artificial entities incorporated in their territory have acquired their nationality. For instance, the *Singapore Deep Sea Mining Act* defines a Singapore company as 'a company incorporated in Singapore.'[47]

However, it must be noted that the ILC identified some possible exceptions to the acquisition of nationality by incorporation in a particular State when there is no evidence of some sort of significant link or connection between the State of incorporation and the corporation itself but rather significant connections with other States, namely situations when the corporation is controlled by nationals of another State or States and has no substantial business activities in the State of incorporation, and the seat of management and the financial control of the corporation are both located in another State. In the latter situation, the other State(s) may be regarded as the State of nationality for the purpose of diplomatic protection.[48] It should be noted that in the case of DSM, State sponsorship may be done either by the State of nationality of the Company or the State that effectively controls such Company or whose nationals have effective control over it.[49] Although the issue of the nationality of corporations may generate some debate in terms of diplomatic protection, the author is doubtful that this would actually be much of a practical problem if a foreign corporation incorporates a company in a developing State for purpose of doing business. Usually the expectation of the developing State will be that such foreign company incorporated under its domestic laws would carry out a substantial amount of business within its territory which therefore would generate some type of economic benefits to such State and its citizens. For example, in the case of deep seabed mining, the Secretariat of the Pacific Community (SPC) in a report prepared in 2012 recognized that there could be huge economic benefits which could accrue to African Caribbean and Pacific (ACP) States that sponsor deep seabed mining exploitation in the Area.[50] The report pointed out that potentially this could benefit ACP States

---

46 See Article 9 and ILC Commentary, Draft Articles on Diplomatic Protection, ILC Report of the 58th Session, 2006.

47 Section 2 of Singapore Deep Sea Mining Act (adopted 6 March 2015, entered into force 1 April 2015).

48 See Article 9 and ILC Commentary, Draft Articles on Diplomatic Protection, ILC Report of the 58th Session, 2006.

49 Art. 153(2) of UNCLOS.

50 Secretariat of the Pacific Community, 'Pacific-ACP States Regional Legislative and Regulatory Framework for Deep Sea Minerals Exploration and Exploitation: Prepared under the SPC-EU EDF10 Deep Sea Minerals Project', <http://www.smenet.org/docs/public/FinalDeepSeaMineralsProjectReport.pdf> accessed 1 July 2017.

by contributing to its government revenues through taxes and/or royalties, by creating jobs and training opportunities as well as strengthening the domestic private sector, by encouraging foreign investment and the funding of public service improvements, and by contributing to infrastructure development and supporting other economic sectors.[51] Usually, when foreign TNCs incorporate subsidiary companies under the laws of developing States, the government insists on either the government having some shares in the company themselves or its citizens being given the opportunity to invest and acquire shares in the company. For instance, under the Nigerian Maritime Administration and Safety (NIMASA) Act 2007, 'Nigerian Company' is defined for the purposes of the legislation as 'a company incorporated in Nigeria in which Nigerian citizens hold at least 60 per cent of the shares as beneficial owners.'[52]

Discussing circumstances where foreign companies incorporate subsidiaries in developing States, the Seabed Chamber, obviously having in mind the issue of ships and flags of convenience, focuses on the so-called Sponsoring States of Convenience situation. It explains this as a situation where commercial enterprises based in developed States setup companies in developing States, thereby acquiring the latter States' nationality and obtaining their sponsorship to mine the Area 'in the hope of being subjected to less burdensome regulations and controls.'[53] The center of attention of the Chamber appears to be on this rather negative perspective of foreign companies establishing subsidiaries in developing States.[54] Whilst this is no doubt a cause for concern, especially with the experiences of how certain TNCs have been guilty of engaging in poor environmental practices when they are involved in mining activities in developing States, including those in Africa, which have lax legislation and regulations.[55] It is important, however, to point out that there may be other motivations, not as sinister as the desire to be Sponsoring States of Convenience, why developing States may consider allowing TNCs to incorporate subsidiaries in their territories to engage in DSM activities in the Area. For one, developing States that do

51 Ibid., p. 6, para. 4.2.

52 Section 64 of Nigerian Maritime Administration and Safety Agency Act (entered into force 25 May 2007), Act No. 17, 2007.

53 ITLOS Case No. 17, para. 159.

54 B.N. Metaxas, 'Flags of Convenience', *Marine Policy*, 1981, pp. 52–66.

55 E. Egede, 'Human Rights and the Environment: Is there a Legally Enforceable Right to a Clean and Healthy Environment for the "Peoples" of the Niger Delta under the Framework of the 1999 Constitution of Nigeria', *Sri Lanka Journal of International Law*, vol. 19, 2007, pp. 51–83.

not have the financial and technical capacity may wish indirectly to engage in deep seabed mining activities by working with TNCs that have such a financial and technical capability. After all, Annex III of the UNCLOS and the provisions of the different regulations make it clear that the 'qualification standards shall relate to the financial and technical capabilities of the applicant and his performance under any previous contracts with the Authority.'[56] This model, which could be called the 'Nauru/Tonga model', would appear to be the way forward to get developing States lacking the requisite financial and technical capability to speedily participate in mining activities in the Area and get on the Seabed mining ladder. This model by itself is nothing new, as we see similar arrangements like this reflected in the way developing States mine natural resources within their national jurisdiction. Nauru in its proposal to the ISA requesting that an advisory opinion be obtained from the Seabed Dispute Chamber of the ITLOS puts it this way:

> Nauru, like many other developing States, does not yet possess the technical and financial capacity to undertake seafloor mining in international waters. To participate effectively in activities in the Area, these States must engage entities in the global private sector (in much the same way as some developing countries require foreign direct investment).[57]

A perusal of Nauru's rather detailed proposal shows clearly that the intention was certainly not to be some sort of sponsoring State of convenience akin to the much maligned flags of convenience States in relation to shipping, but rather a desire to be a responsible developing DSM State engaged in the CHM and working with foreign TNCs with the financial and technical capabilities. In its proposal to the ISA, Nauru pointed out that the idea of its sponsorship of Nauru Oceans Resources Inc. was initially premised on the supposition that:

---

56 Art. 4(2) of Annex III of UNCLOS and Regulation 12(1) of the Regulations on Prospecting and Exploration for Polymetallic Nodules in the Area; Regulation 13(1) of the Regulations on Prospecting and Exploration for Polymetallic Sulphides in the Area; Regulations on Prospecting and Exploration for Cobalt-rich Ferromanganese Crusts in the Area; for further information see footnotes 6 to 8.

57 ISA, Proposal to Seek an Advisory Opinion from the Seabed Disputes Chambers of the International Tribunal for the Law of the Sea on Matters Regarding Sponsoring State Responsibility and Liability (submitted by the delegation of Nauru), ISBA/16/C/6 (5 March 2010).

> Nauru could effectively mitigate (with a high degree of certainty) the potential liabilities or costs arising from its sponsorship. This was important, as these liabilities or costs could, in some circumstances, far exceed the financial capacities of Nauru (as well as those of many other developing States). [...] Nauru [thus] considers it crucial that guidance be provided on the interpretation of the relevant section of Part XI pertaining to responsibility and liability, so that developing States can assess whether it is within their capabilities to effectively mitigate such risks and in turn to make an informed decision on whether or not to participate in activities in the Area.[58]

The Nauru/Tonga model must therefore be distinguished from the Sponsored States of convenience model. The idea of the Nauru/Tonga model is to provide a platform for developing States that do not presently have the financial/technical capability to engage with deep seabed mining activities through the use of the time honored practice of encouraging Foreign Direct Investment (FDI) in developing States. Surely, this could be said to be in line with Article 148 of the UNCLOS which states:

> The effective participation of developing States in activities in the Area shall be promoted as specifically provided for in this Part, having due regard to their special interests and needs, and in particular to the special need of the land-locked and geographically disadvantaged among them to overcome obstacles arising from their disadvantaged location, including remoteness from the Area and difficulty of access to and from it.

For the TNCs, on the other hand, the motivation for incorporating companies in developing States may not necessarily be the desire to avoid burdensome regulations, but rather the incentive pull of access to the reserved areas. The reserved area would obviously be more cost effective because the TNC would not have to submit an application that would cover an area sufficiently large and of sufficient estimated commercial value to allow for two mining operations.[59] As the UNCLOS states:

> Any State Party which is a developing State or any natural or juridical person sponsored by it and effectively controlled by it or by other developing State which is a qualified applicant, or any group of the foregoing, may

58 Ibid., para. 1 and 5.

59 Art. 8 of Annex III of UNCLOS; see also Section 1 para. 10 of the Annex to the Agreement.

> notify the Authority that it wishes to submit a plan of work pursuant to Article 6 of this Annex with respect to a reserved area. The plan of work shall be considered if the Enterprise decides, pursuant to paragraph 1, that it does not intend to carry out activities in that area.[60]

Also Regulation 17 of the Regulations on Prospecting and Exploration for Polymetallic Nodules in the Area, for instance, states:

> Any State which is a developing State or any natural or juridical person sponsored by it and effectively controlled by it or by any other developing State, or any group of the foregoing, may notify the Authority that it wishes to submit a plan of work for exploration with respect to a reserved area. The Secretary-General shall forward such notification to the Enterprise, which shall inform the Secretary-General in writing within six months whether or not it intends to carry out activities in that area. If the Enterprise intends to carry out activities in that area, it shall, pursuant to paragraph 4, also inform in writing the contractor whose application for approval of a plan of work for exploration originally included that area.

4 Seabed Chamber Advisory Opinion: Responsibilities and Obligations of Developing States

It would be recalled that what eventually led to the ITLOS Seabed Chamber's advisory opinion was actually elicited by the proposal of Nauru, a developing State, which eventually led to the Council of the ISA reformulating and seeking the advisory opinion of the Chamber on the following three specific questions:[61]

- What are the legal responsibilities and obligations of States Parties to the Convention with respect to the sponsorship of activities in the Area in accordance with the Convention, in particular Part XI, and the 1994 Agreement relating to the implementation of Part XI of UNCLOS 1982?

60 See also Section 2, para. 5 of the Annex to the Agreement.

61 ISA, Proposal to Seek an Advisory Opinion from the Seabed Disputes Chambers of the International Tribunal for the Law of the Sea on Matters Regarding Sponsoring State Responsibility and Liability (submitted by the delegation of Nauru), UN Doc. ISBA/16/C/6 (5 March 2010).

- What is the extent of liability of a State Party for any failure to comply with the provisions of the Convention, in particular Part XI, and the 1994 Agreement, by an entity whom it has sponsored under Article 153, paragraph 2 (b), of the Convention?
- What are the necessary and appropriate measures that a sponsoring State must take in order to fulfil its responsibility under the Convention, in particular Article 139 and Annex III, and the 1994 Agreement?[62]

The Chamber in answering these questions was clear about the need for equality of treatment between developing and developed sponsoring States under the UNCLOS as regards responsibilities and liabilities with respect to the protection of the marine environment so as not to 'jeopardise uniform application of the highest standards of protection of the marine environment, the safe development of activities in the Area and protection of the common heritage of mankind.'[63] In essence, under the UNCLOS the requirement to take the needs and interests of developing States into consideration (in some regards in Part XI), did not in itself permit a lower standard of protection of the marine environment in the Area.[64] Thus the principle of common but differentiated responsibilities enshrined International Environmental Law was not applicable because it was not specifically stated in the UNCLOS. In response to the questions put before it by the Council, the Chamber identified that all sponsoring States, both developing and developed, have an obligation (responsibility) to ensure that the activities in the Area conducted by entities they have sponsored are in conformity and in compliance with the UNCLOS, the 1994 Agreement and the regulations, rules and procedures of the ISA and direct obligations.[65] The sponsoring State Party also has direct obligations, identified by the Chamber, as obligations that sponsoring States 'have to comply independently of their obligation to ensure a certain behavior by the sponsored contractor.'[66] The Chamber points out that the most important of these direct obligations are: the obligation to assist the ISA in the exercise of control over activities in the Area; the obligation to apply a precautionary approach; the obligation to apply best environmental practices; the obligation to take measures to ensure the provision of guarantees in the event of an emergency order by the ISA for protection of the marine environment; the obligation to ensure

62 ITLOS Case No. 17, para. 1.

63 Ibid., para. 158, 159.

64 See Articles 140, 160(f) of UNCLOS.

65 Para. 103, 104 and 121.

66 Para. 121.

the availability of recourse for compensation in respect of damage caused by pollution; and the obligation to conduct environmental impact assessments. Although the obligations to ensure are different than the direct obligations, the Chamber was quick to point out the linkage between the two since compliance with the direct obligations may be regarded as a relevant factor in meeting the obligation to ensure.[67]

The Chamber pointed out that a violation of the obligations leads to liability. However, it also pointed out that not every violation of an obligation by the sponsored contractor would necessarily lead to the liability of the sponsoring State. In essence, there is no strict liability on the sponsoring State. There had to be a proven causal link between the failure of the sponsoring State to carry out its obligation and the damages caused by the sponsored entity.[68] The obligation is thus one of conduct and not of result amounting to one of 'due diligence.'[69] This 'due diligence' obligation, in essence, requires sponsoring States Parties to 'make best possible efforts' to secure the compliance by sponsored entities with their obligations by adopting the appropriate regulatory and administrative measures within their domestic legal system and to take appropriate steps in their enforcement.[70] The obligation is not to ensure that the contractors always carry out their own obligations but for sponsoring States to take these measures within their legal system that are 'necessary and appropriate' to ensure that contractors discharge these obligations.[71] Although the Chamber did not give an exhaustive list of what should be done to fulfil this obligation to ensure, as this was outside its remit, it did give some directions in this regard.[72] It identified as essential the need to have appropriate domestic legislation, regulations, and administrative measures, including the establishment of enforcement mechanisms to actively supervise the activities of the sponsored contractors and provide for some sort of coordination between the sponsoring State and the ISA.[73] It pointed out that the existence of the appropriate laws, regulations and administrative measures in the domestic legal system of a sponsoring State is 'a necessary requirement for compliance with the obligation of due diligence of the sponsoring State and

67 Para. 122 and 124–150.

68 ITLOS Case No., para. 109–110, 172–173, 181 and 189.

69 The Chambers referred extensively to the ICJ decision in Pulp Mills on the River Uruguay (Argentina v. Uruguay), Judgement, ICJ Reports 2010, p. 14.

70 Para. 242.

71 Ibid.

72 Para. 227.

73 Para. 218.

for its exemption from liability.'[74] It indicated that if a Sponsoring State merely enters into a contractual arrangement with the sponsored entity without appropriate domestic legislation regulating DSM, it would not be regarded as complying with its due diligence obligation.[75]

Over the years certain developing States have adopted Deep Seabed Mining legislation, some more detailed than others, especially regarding effective protection of the marine environment against any harmful effects of DSM activities. For example the Singapore Deep Seabed Mining Act 2015 was enacted to regulate activities in the Area by persons sponsored by Singapore; to ensure the effective protection of the marine environment against any harmful effects of those activities and to fulfil Singapore's obligations under the UNCLOS and the 1994 Agreement with regards to such activities.[76] The legislation indicates a preference for sponsoring Singaporean companies rather than individuals.[77] An individual who is a Singapore citizen may only explore or exploit a resource in the Area if such individual is acting as an employee or agent of a Singapore company that has an existing licence and ISA contract, or a person who has entered into an ISA contract, which remains in force, under the sponsorship of another State, and such individual's exploration or exploitation activities are within the scope of the licence granted to such Singapore company, that the individual is an employee or agent of, or the existing ISA contract that has been sponsored by such other State.[78] It states a Singaporean company may explore for or exploit a resource in a part of the Area if the company has been granted a licence, which remains in force to do so, and the company has entered into a contract with the ISA. It makes provision for the following consequences: a general prohibition of unilateral deep seabed mining making the contravention of this an offence; a licensing regime whereby a Singaporean company wishing to be sponsored by Singapore had to apply for a licence from the appropriate minister and the conditions for granting a sponsorship certificate; enforcement of the Seabed Disputes Chamber's decisions; identifies some miscellaneous offences and powers given by the minister to make regulations pursuant to the Act. The legislation requires an entity that wishes to engage in mining activities to first apply for a licence. It is only when the licence has been granted by the appropriate minister that Singapore may sponsor the licencee's

74 Para. 219.

75 Para. 223–226.

76 Agreement Relating to the Implementation of Part XI of the United Nations Convention on the Law of the Sea of 10 December 1982 (28 July 1994) (1994 Agreement), Section 3.

77 Ibid., Section 5(1) and 7.

78 Ibid., Section 5(2).

application to the ISA and issue a certificate of sponsorship.[79] The Singaporean legislation is however sparse in engaging with measures to be taken to protect the environment from such mining activities. Reference to protecting the environment is scantly stated in the section of the Act mentioning that one of the purposes of the legislation is 'to effectively protect the marine environment against any harmful effects of those activities or the cessation of those activities.'[80] The only other reference to protecting the environment from DSM activities in other parts of the legislation is in relation to the directions of the appropriate minister when a licence expires etc., which mentions that the relevant minister may give directions 'that specified measures be taken to effectively prevent, contain or minimize any harmful effects to the marine environment.'[81] It however does not explicitly refer to obligations, referred to by the ITLOS Chamber, such as the need to apply a precautionary approach; best environmental practices; the need to take measures to ensure the provision of guarantees in the event of an emergency order by the ISA for protection of the marine environment; the need to ensure the availability of recourse for compensation in respect of damage caused by pollution; and the need to conduct environmental impact assessments. It also does not establish specific institutions, apart from reference to the appropriate minister, with enforcement powers to regulate and monitor domestically the sponsored contractors. Perhaps the idea is for the relevant minster to spell this out in subsequent Regulations which the minister certainly has the powers to make under the Act.[82]

On the other hand, the legislation of Island Pacific States, such as Fiji,[83] Nauru,[84] Tonga[85] and Tuvalu[86] appear to be more detailed, especially as it regards their engagement with measures to ensure the protection of the marine environment as a result of DSM activities. These States have benefited from the Pacific Regional Legislative and Regulatory Framework (RLRF), which was made available by the Secretariat of the Pacific Community (SPC) to guide Pacific Island developing States with regard to preparing the appropriate national legislative and regulatory framework in seabed mining to draft na-

79 Ibid., Sections 4, 6–8.

80 Ibid., Section 3(b).

81 Ibid., Section 16(2)(b) and 16(4).

82 Section 24.

83 Fiji International Seabed Mineral Management Decree, No. 21 of 2013.

84 Nauru International Seabed Minerals Act, No. 26 of 2015.

85 Tonga Seabed Minerals Act, No. 10 of 2014.

86 Tuvalu Seabed Minerals Act, No. 14 of 2014.

tional legislation. The framework which is a rather comprehensive document, put together with the assistance of several interested stakeholders, provides the Pacific Island States with the tools to put together a comprehensive national policy, legal framework and institutional capacity.[87] According to Makgill and Linhares, the 'Advisory Opinion, therefore, provided strong incentive for Pacific States, entertaining seabed mining activities within their jurisdiction to enact appropriate laws and regulation. Consequently the precautionary approach, best environmental practices, and the EIA have all found expression in a number of subsequent legislative initiatives across the region.'[88]

These legislations are clearer in emphasizing the role of the sponsoring States to exercise effective control over the sponsored person, as well as their responsibilities, especially as it concerns the protection of the marine environment from DSM activities through the application of the precautionary approach and best environmental practices than that of Singapore.[89] For instance, the Fiji International Seabed Mineral Management Decree states that any person engaged in Seabed Mineral activities whether directly or indirectly shall, amongst other things, be required to 'apply the Precautionary Approach, and employ best environmental practice in accordance with prevailing international standards in order to avoid, mitigate or remedy adverse effects of Seabed Minerals Activities on the marine environment.'[90] Also, the Seabed Minerals Act 2014 of Tonga is clearer on the responsibility of the sponsored entity with regard to the protection of the environment. For instance it states that: 'The terms of any environmental conditions arising from an Environmental Impact Assessment conducted in compliance with the EIA Act shall be adopted as part of the terms and conditions of any title held under this Act.'[91] Furthermore, the legislation of these Pacific Island States provide for the establishment of some sort of domestic regulatory and monitoring body to ensure the effective implementation of their respective legislation, includ-

87 See <http://dsm.gsd.spc.int/> accessed 1 July 2017; and M. Borrel, 'Protection and Preservation of the Marine Environment from Seabed Mining Activities on the Continental Shelf: Perspectives from the Pacific Islands Region', in R. Rayfuse (ed.), *Research Handbook on International Marine Environment Law*, Edward Elgar, 2015, pp. 206–228, esp. pp. 210–211.

88 R. Makgill and A.P. Linhares, 'Deep Seabed Mining: Key Obligations in the Emerging Regulation of Exploration and Development in the Pacific', in R. Warner and S. Kaye (eds.), *Routledge Handbook of Maritime Regulation and Enforcement*, Routledge, 2015, pp. 231–261 (232).

89 See for instance Section 30 of the Nauru International Seabed Minerals Act, No. 26 of 2015.

90 Section 32(e).

91 Section 102.

ing the protection of the environment. An example of this is the Tonga Seabed Minerals Authority, which was created amongst other things to:

- review or obtain a review of Environmental Impact Assessments for Seabed Mineral Activities required under this Act and the EIA Act;
- liaise with the ISA and any other relevant international organization in accordance with the UN Convention of the Law of the Sea to facilitate the lawful conduct of Seabed Minerals Activities or the protection of the Marine Environment;
- seek expert advice on factual matters pertaining to the administration of this Act and concerning the management of the Kingdom's Seabed Minerals, including but not limited to advice on economic, legal, scientific, and technical matters, and including advice from experts in the management and conservation of the Marine Environment.[92]

The point that obviously comes out in exploring the various domestic legislations of developing States is the need for some kind of consistency in legislation on DSM, especially as regards the Sponsoring States taking effectual control under domestic legislation to regulate and monitor sponsored entities so as to ensure they adopt the best environmental practices. Clearly, there is a need to pursue some type of harmonization of national legislation of sponsoring States, especially those from developing States. One of the ways to achieve such harmonization that would ensure some consistency and compliance with minimum international standards is through the utilization of model legislation.[93] The ISA could develop model DSM legislation that would contain detailed relevant clauses that would provide the appropriate legal framework for sponsoring States to monitor and regulate their sponsored entities. As pointed out by the ITLOS Chamber, laws and regulations are not in themselves enough to exclude the sponsoring State's liability but also there is a need for administrative measures, including the establishment of enforcement mechanisms to ensure effective supervision of the sponsored contractors.[94] Thus, apart from such model legislation, there would be a need for the ISA to be involved in capacity building to assist sponsoring States, especially developing States, to develop an effective regulatory/enforcement body composed of well-trained personnel to ensure that sponsored entities comply

---

92 Section 12.

93 See G. Kent, 'Harmonizing Extended Zone Legislation in Southeast Asia', *Ocean Development & International Law*, vol. 13, 1983, pp. 247–268 (250–254).

94 See para. 218.

with minimum international standards, including standards on the protection of the environment. This could be done by requiring States that have sponsored entities to send some of the key personnel of their DSM regulatory bodies for periodic training under the auspices of the ISA, which would be regarded as International Continuing Professional Development (ICPD), to familiarize them with the best international environmental practices in respect of DSM.

5 Now where are African Developing States in DSM Activities?

At present Africa is the only region, utilizing the International Seabed Authority regional groupings,[95] which does not have States that have sponsored contractors to engage in DSM activities.[96] The 2050 Africa's Integrated Maritime (AIM) Strategy, a long-term strategy aimed at providing a broad framework for the protection and sustainable exploitation of the African Maritime Domain (AMD), adopted by the African Union in January 2014,[97] failed to specifically consider the Deep Seabed Area beyond National Jurisdiction (the Area) and the resources therein. Neither does it mention the possibility of the Continent engaging with DSM activities in the Area.[98] It is rather astonishing that the AIM Strategy, intended to be a long term strategy, is silent on the Area and DSM activities, especially since Africa played a key role at the Third United Nations Conference on the Law of the Sea (UNCLOS III) in the development of the regime of the deep seabed beyond national jurisdiction (the Area), which was eventually adopted as Part XI of the UNCLOS.[99] Furthermore, there is a strong linkage between the Continental Shelf (CS) and the Area as the outer limits of the CS of certain coastal States are used to delineate the Area.[100] Thus the outer limits of the Continental Shelf of some African States facing either the Indian or the Atlantic Ocean actually act as boundaries for the Area. Consequently, the DSM activities adjacent in the Indian and Atlantic Oceans would have some implications for these States, including environmental

95 See <https://www.isa.org.jm/regional-groups> accessed 1 July 2017; these regional groupings are in line with United Nations geographical grouping for election purposes.

96 <https://www.isa.org.jm/deep-seabed-minerals-contractors> accessed 1 July 2017.

97 <http://pages.au.int/maritime/documents/2050-aim-strategy-0> accessed 1 July 2017.

98 For analysis of this see E. Egede, 'Institutional Gaps in the 2050 Africa's Integrated Maritime Strategy', *Journal of Ocean Law & Governance in Africa*, vol. 1, 2016, pp. 1–27 (19–26).

99 See Declaration of the Organization of African Unity on the Issues of the Law of the Sea (adopted 19 July 1974), UN Doc. A/CONF.62/33 (1974), esp. para. 18–22.

100 Article 1(1) (1) of UNCLOS 1982.

impacts. Moreover, certain mineral resources may actually straddle between the Area and the adjacent Continental Shelves of certain African States and thus be of direct interest to these States.[101]

However, currently there appears to be a growing appetite for Africa to begin to engage with DSM. For instance, the final edition of an African Union document setting out the agenda for 2063, under its Aspiration 1 titled 'A Prosperous Africa based on inclusive growth and sustainable development', includes the following:

> Africa's Blue economy which is three times the size of its landmass, shall be a major contributor to continental transformation and growth, advancing knowledge on marine and aquatic biotechnology, the growth of an Africa-wide shipping industry, the development of sea, river and lake transport and fishing; *and exploitation and beneficiation of deep sea mineral and other resources.*[102]

Furthermore, under Aspiration 7 of this document it stated: '[w]e aspire that by 2063, Africa shall be: [a] major social, political and economic force in the world, and with her rightful share of the global commons (land, oceans and space) [...].'[103]

A swift way by which African States may participate in DSM is by adopting the Tonga/Nauru model,[104] the approach adopted by certain Pacific States.

---

101 Article 142 of UNCLOS 1982.

102 Aspiration 1, para. 15 of Agenda 2063: The Africa We Want, September 2015, popular version (Agenda 2063), <http://agenda2063.au.int/en/sites/default/files/03_Agenda2063_popular_version_ENG%2021SEP15-3.pdf> accessed 1 July 2017 (emphasized by the author). See also the Framework Document: The Africa We Want of Agenda 2063, September 2015, which at page 103 identifies the need to: '[d]evelop capacities to access global commons in areas beyond national jurisdiction and deep seas.' <http://agenda2063.au.int/en/sites/default/files/Framework%20Document%20Book.pdf> accessed 1 July 2017. Also see references at pp. 114, 127 of this Framework Document to Deep Seabed Mining.

103 Aspiration 7, para. 60 of Agenda 2063.

104 See however E. Egede, *Africa and the Deep Seabed Regime: Politics and International Law of the Common Heritage of Mankind*, Springer, 2011, Chapter 7, where the author proposes different strategic alliances and cooperative efforts that Africa may employ to achieve DSM in the long run. Also, see E. Egede, 'Institutional Gaps in the 2050 Africa's Integrated Maritime Strategy', *Journal of Ocean Law & Governance in Africa*, vol. 1, 2016, pp. 1–27 (24) where the author advocates various interested African States pooling their resources under the auspices of the African Union to put together some type of African Deep Seabed Mining Corporation.

For instance, Nautilus Minerals Inc., a TNC, incorporated local subsidiaries, namely Nauru Ocean Resources Inc. and Tonga Offshore Mining Limited, in the Republic of Nauru and the Kingdom of Tonga respectively, and the local companies were sponsored by the State where they have been incorporated for work plans for explorations in the part of the Area reserved by the ISA for developing countries.[105] It is suggested that such African States could further explore the option of acquiring some equity interest in such subsidiary corporations and also encouraging their nationals to do so, if they wish.

A key initial measure interested African States would need to put in place is the appropriate domestic legal framework to ensure they exercise effective regulatory oversight over entities they would sponsor to engage in DSM activities, so as to be exempt from liability as a result of such entity's failure to carry out its responsibilities under the UNCLOS and relevant Regulations, as well as to safeguard against being labelled as sponsoring States of Convenience, as elucidated by Advisory Opinion of the ITLOS Chamber. Such legislation should be 'no less effective than international rules, regulations and procedures' such as the UNCLOS and the ISA Mining Codes. Such legislation should also, amongst other things, certify that such States are ensuring that their sponsored contractors are carrying out DSM activities in line with best environmental practice, including adopting the precautionary approach and conducting effective prior environmental impact assessment. It should be noted that the need to adopt the precautionary approach as it regards environmental protection is already reflected in certain treaties adopted by the African Union. For instance, the *Bamako Convention on the Ban of the Import of Hazardous Wastes into Africa and on the Control of their Transboundary Movements within Africa*,[106] states in Article 4(f) that:

> Each party shall strive to adopt and implement the preventive, precautionary approach to pollution problems which entails, inter alia, preventing the release into the environment of substances which may cause harm to humans or the environment without waiting for scientific proof regarding such harm. The Parties shall co-operate with each other in

105 Available at <http://www.isa.org.jm/files/documents/EN/Press/Press11/SB-17–11.pdf> accessed 1 July 2017.

106 Bamako Convention on the Ban of the Import of Hazardous Wastes into Africa and on the Control of their Transboundary Movements within Africa (adopted 29 January 1991, entered into force 22 April 1998) 30 ILM 773 (1991) (Bamako Convention). This treaty as at 24 October 2016 has been ratified by 25 out of 54 AU members. <http://www.au.int/en/sites/default/files/treaties/7774-sl-bamako_convention.pdf> accessed 1 July 2017.

> taking the appropriate measures to implement the precautionary principle to pollution prevention through the application of clean production methods, rather than the pursuit of a permissible emissions approach based on assimilative capacity assumptions.

While the African Convention on the Conservation of Nature and Natural Resources, as revised in 2003, in Article IV states:[107]

> The Parties shall adopt and implement all measures necessary to achieve the objectives of this Convention, in particular through preventive measures and the application of the precautionary principle, and with due regard to ethical and traditional values as well as scientific knowledge in the interest of present and future generations.

Concerning the direct obligation of sponsoring States to ensure that EIAs are conducted, which the ITLOS Chamber found to extend beyond the scope of the ISA regulations and was a direct obligation under the Convention and Customary International Law,[108] there is evidence that African States are already engaging with the issue of the need to conduct EIAs as regard to projects that have environmental implications. A review report by the United Nations Economic Commission for Africa (UNECA), as far back as 1995,[109] revealed that out of 23 African States reviewed, 18 of them had an enabling legislation and/or specific legislation on EIA, with ten of these States actually having explicit provisions for public participation. However, the report pointed out certain problems with effective implementation of EIA in Africa such as: the challenge

---

107 See African Convention on the Conservation of Nature and Natural Resources (adopted 11 July 2003) (Maputo Convention). This treaty as at 24 October 2016 had been ratified by 13 out of 54 AU members. <http://www.au.int/en/sites/default/files/treaties/7782-sl-revised_-_nature_and_natural_resources_1.pdf> accessed 1 July 2017.

108 Paras. 145 and 148 of Maputo Convention.

109 Economic Commission for Africa (ECA), *Review of the Application of Environmental Impact Assessment in Selected African Countries*, 1995, <http://www1.uneca.org/Portals/3/documents/EIA_book_final_sm.pdf> accessed 1 July 2017. For example, the Nigeria Environmental Impact Assessment Act No. 86 of 1992 was enacted in response to the outcome of the United Nations Conference on Environment and Development (UNCED) held in Rio de Janeiro in 1992. It outlines the goals and objectives of an EIA, the minimum content of an EIA and a list of activities that are not permitted to go ahead until the Agency established under the legislation has been consulted and has given its approval. The main goal of the Act is to ensure environmentally sound and sustainable development projects.

of operationalizing the institutional and regulatory frameworks for EIA; inter-agency collaboration and decentralization; EIA capacity issues; the quality of the EIA studies and review processes; EIA costs and timeframes; public participation; implementation of the Environmental Management Plan (EMP) & follow-up; integration of the EIA system into an overall Environmental Management System (EMS).[110] This therefore indicates the need for any such African State wishing to engage in sponsoring contractors for DSM activities to ensure that not only the appropriate national DSM legislation/regulation is in place, but also that the appropriate regulatory/enforcement bodies, manned by qualified personnel, are established within domestic jurisdiction to undertake the effectual monitoring/regulation of sponsored contractors. The Pacific RLRF, subject to adjustments to suit any African specific situation, may be used as a model to assist African States interested in DSM to draft and adopt appropriate national legislation to regulate DSM and to meet obligations under the UNCLOS.

6 Conclusion

There is no clear-cut definition of what States fall under the category of developing States under the UNCLOS, as amended by the 1994 Agreement, and the ISA mining Code. The phrase by its nature is a political one that would eschew any effort to come up with a precise legal definition. However, what is clear is that there is a 'mixed bag' of developing States out in the world backdrop, with some more developed than others. This is reflected in DSM activities where you have some developing States already engaged with such activities while others are yet, so to speak, to climb unto the DSM 'ladder'. The Advisory Opinion of the ITLOS Chamber made it clear, however, that while Part XI of the UNCLOS, as amended, gives certain preferential treatment to encourage developing States to participate in DSM activities, no such special treatment was given to such States regarding the general obligation to protect the marine environment from DSM. The Chamber pointed out that the highest environmental standards were required by all States, both developed and developing, with regard to DSM activities to avoid the emergence of the phenomenon of Sponsoring States of Convenience. The Chamber emphasized that it was vital for Sponsoring States, including developing States, to put in place the appropriate legislative and institutional framework to ensure such States

110 See ibid.

effectually regulated and monitored their Sponsored contractors to ensure that they maintain the highest environmental standards in carrying out DSM activities. However, a survey of some of the current DSM legislation reveals a lack of harmonization, with some being more comprehensive than others. Developing States, such as Pacific Island States, involved in DSM activities have a reasonably comprehensive legislative and institutional framework, which is harmonized under the auspices of the SPC, and could provide a good model for other developing States, especially those from Africa, who may wish to engage in DSM. To achieve some level of consistency in legislative and institutional framework this chapter has proposed the need for the ISA to put together a comprehensive model legislation that States interested in DSM may wish to adopt as their national legislation, with necessary adjustments to suit their individual domestic needs, and also for the ISA to put together regular training on best environmental practices, especially for personnel of the DSM regulatory bodies in developing States, so they may be equipped to operationalize said practices in their respective States.

Bibliography

Anton, D.K. 'The Principle of Residual Liability in the Seabed Disputes Chamber of the International Tribunal for the Law of the Sea: The Advisory Opinion on Responsibility and Liability for International Seabed Mining (ITLOS Case No. 17)', *McGill Int'l J. Sust. Dev. L. & Pol'y*, vol. 7, 2011/2012, pp. 241–257.

Anton, D.K., Makgill, R.A. and Payne, C.R., 'Seabed Mining – Advisory Opinion on Responsibility and Liability', *Environmental Policy and Law*, vol. 41, 2011, pp. 60–65.

Brown, E.D., 'The 1994 Agreement on the Implementation of Part XI of the UN Convention on the Law of the Sea: Breakthrough to Universality?', *Marine Policy*, vol. 19, 1995, pp. 5–20.

Egede, E., 'Human Rights and the Environment: Is there a Legally Enforceable Right to a Clean and Healthy Environment for the "Peoples" of the Niger Delta under the framework of the 1999 Constitution of Nigeria', *Sri Lanka Journal of International Law*, vol. 19, 2007, pp. 51–83.

Egede, E., *Africa and the Deep Seabed Regime: Politics and International Law of the Common Heritage of Mankind*, Springer, 2011.

Egede, E., 'Institutional Gaps in the 2050 Africa's Integrated Maritime Strategy', *Journal of Ocean Law & Governance in Africa*, vol. 1, 2016, pp. 1–27.

Freestone, D., 'Responsibilities and Obligations of State Sponsoring Persons and Entities With Respect to Activities in the Area', *American Journal of International Law*, vol. 105, 2011, pp. 755–760.

French, D., 'From the Depths: Rich Pickings of Principles of Sustainable Development and General International Law on the Ocean Floor – the Seabed Disputes Chamber's 2011 Advisory Opinion', *International Journal of Marine and Coastal Law*, vol. 26, 2011, pp. 525–568.

Kent, G., 'Harmonizing Extended Zone Legislation in Southeast Asia', *Ocean Development & International Law*, vol. 13, 1983, pp. 247–268.

Mero, J.L., *The Mineral Resources of the Sea*, Elsevier Publishing, 1968.

Metaxas, B.N., 'Flags of Convenience', *Marine Policy*, vol. 5, 1981, pp. 52–66.

Oppenheim, L., *International Law: A Treatise*, Vol. 1, Longmans, 1937.

Rayfuse, R. (ed.), *Research Handbook on International Marine Environment Law*, Edward Elgar, 2015.

Sands, P. et al., *Principles of International Environmental Law*, 3rd Edition, Cambridge University Press, 2012.

Sullivan, A. and Sheffrin, S.M., *Economics: Principles in Action*, Prentice Hall, 2003.

Tanaka, Y., 'Obligations and Liability of Sponsoring States Concerning Activities in the Area: Reflections on the ITLOS Advisory Opinion of 1 February 2011', *Netherlands International Law Review*, vol. 60, 2003, pp. 205–230.

Warner, R. and Kaye, S. (eds.), *Routledge Handbook of Maritime Regulation and Enforcement*, Routledge, 2015.

CHAPTER 9

State Practice in Deep Seabed Mining: The Case of the People's Republic of China

Keyuan Zou

1 Introduction

Under the 1982 United Nations Convention on the Law of the Sea (UNCLOS), the international deep seabed area is named "the Area", which, together with its resources, is the common heritage of mankind.[1] No State should claim or exercise sovereignty or sovereign rights over any part of the Area or its resources, nor should any State or natural or legal person appropriate any part thereof.[2]

Since deep seabed areas beyond the limit of national jurisdiction have been treated as the common heritage of mankind, all activities there are governed by the international regime created under the UNCLOS. The International Seabed Authority (ISA) was accordingly established on 16 November 1994, upon the entry into force of the UNCLOS. The ISA has its headquarters in Kingston, Jamaica and functions as the representative of the whole of mankind for the management of deep seabed mining. The ISA is an autonomous international organization under the UNCLOS and the 1994 Agreement relating to the Implementation of Part XI of the United Nations Convention on the Law of the Sea (the 1994 Agreement),[3] and through it States Parties to the UNCLOS can, in accordance with the regime for the seabed and ocean floor and subsoil thereof beyond the limits of national jurisdiction (the Area) established in Part XI and the 1994 Agreement, organize and control activities in the Area, particularly

* Some parts of this chapter are drawn from my previous article: 'China's Efforts in Deep Seabed Mining: Law and Practice', *International Journal of Marine and Coastal Law*, Vol. 18, 2003, 481–508.

1 United Nations Convention on the Law of the Sea (adopted 10 December 1982, entered into force 16 November 1994) 21 ILM 1261 (1982) (UNCLOS), Art. 136.

2 UNCLOS, Art. 137 (1).

3 Text is available at <http://www.un.org/Depts/los/convention_agreements/texts/unclos/closindxAgree.htm> accessed 4 September 2017.

 | DOI 10.1163/9789004360273_010

with a view to administering the resources of the Area.[4] Deep seabed mining, once described as 'a remote possibility',[5] has gradually become a more realistic proposition, as can be seen from the increased activities carried out by States concerned. In State practice, China's activities and contributions to the deep seabed mining regime merits closer examination.

2 China at UNCLOS III

In 1971, the People's Republic of China (PRC) took the seat in the United Nations as the sole legitimate government of the whole of China, and accordingly sent delegations to attend UNCLOS III from beginning to end (1973–1982). During UNCLOS III, China submitted a working paper specifically on 'General Principles for the International Sea Area'. According to the Chinese view, the ocean and the sea should be divided into two parts under the law of the sea: the sea area within the limits of national jurisdiction, and the international sea area including 'all the sea and ocean space beyond the limits of national jurisdiction'.[6] The international sea area and its resources are, in principle, jointly owned by the people of all countries.[7] It is clear that China's concept of the "international sea area" is different from that of the "international seabed area" in the sense that the Chinese concept does not limit itself to the deep seabed.

China did not use the term "high seas" in referring to the international sea area, as China viewed the term as obsolete, appearing as it did in the four 1958 Geneva Conventions to which China had not acceded: in China's eyes, the Geneva Conventions were representative of the old law of the sea. While the old law of the sea served only the interests of a few big powers, the new UNCLOS laid down a number of important legal principles for safeguarding

4 See 'About the Authority', available at <http://www.isa.org.jm/en/default.htm> accessed 4 September 2017.

5 M. Lodge, 'International Seabed Authority's Regulation on Prospecting and Exploration for Polymetallic Nodules in the Area', *Journal of Energy & Natural Resources Law*, vol. 20, no. 3, 2002, p. 294.

6 'Working Paper on General Principles for the International Sea Area', A/AC.138/SC.II/L.45, 6 August 1973, reprinted in J. Greenfield, *China's Practice in the Law of the Sea*, Clarendon, 1992, p. 235.

7 Ibid.

the common heritage of mankind and the legitimate maritime rights and interests of all States.[8]

In 1972, the Chinese delegation expressed the view that the international regime should not limit its regulation to seabed exploration and exploitation activities. If the regulatory scope of the international seabed area were only limited to the development of seabed mineral resources, it would neither accommodate the interests of developing countries, nor satisfy the concept of the common heritage of mankind.[9] Later on, when the international regime for the deep seabed was finally established, China shifted its original position so as to support the new regime encompassing only the deep seabed.

In criticizing the superpowers, China considered the 'parallel system of exploitation' put forward by one superpower to be a bad idea.[10] Later on, China changed its stance and did not oppose the establishment of the parallel system, both as a compromise and as a provisional arrangement, provided that the necessary technology and resources for the Authority and the Enterprise were guaranteed.[11]

The Chinese delegation was dissatisfied with the final arrangement for the legal regime of deep seabed mining:

> Resolution II of the Conference, governing preparatory investment in pioneer activities relating to polymetallic nodules, has done too much in the way of meeting the demands of a few industrialized nations and given them and their companies some privileges and priorities. We consider that inappropriate.[12]

---

8 Mr. Han Xu, 191st Meeting, Plenary Meetings, 9 December 1982, in United Nations, *Third United Nations Conference on the Law of the Sea: Official Records*, Vol. XVII, 1984, at 102.

9 X. Pu on Ocean International Regime, 27 July 1972, reprinted in Peking University Law Department (ed.), *Collected Materials on the Law of the Sea*, People's Press, 1974, (in Chinese), pp. 35–36.

10 See Mr. Lin Ching, 76th Meeting of the Plenary Meetings, 17 September 1976, in United Nations, *Third United Nations Conference on the Law of the Sea: Official Records*, Vol. VI, 1977, at 26.

11 See Mr. Ke Zaishuo, 114th Meeting of the Plenary Meetings, 26 April 1979, in United Nations, *Third United Nations Conference on the Law of the Sea: Official Records*, Vol. XI, 1980, at 22.

12 Mr. Han Xu, *supra* note 8, at 102.

In the early 1970s, China had no intention of exploring deep seabed mineral resources itself, and had no capability to carry out any deep seabed exploration even if it had had the intention of doing so, as during that time its economy was on the verge of collapse. Only after China had carried out its economic reforms and introduced its "open-door" policy in 1978 did this become feasible.

3 Operational Mechanism

China began its investigation of seabed mineral resources at the end of the 1970s. In 1983, 1985, and 1987, two scientific vessels – *Xiangyanghong 16* and *Xiangyanghong 9* – completed comprehensive surveys including meteorology, hydrology, physical sampling, environment, and geology in the central, north and northwest Pacific. In 1982 China sent a letter to the chairman of the UNCLOS III stating that it had spent 80 million RMB (ca. $US40 million) on polymetallic nodule investigations, but China did not ask to be listed as a potential pioneer investor. After various surveys, an application area totalling 300,000 sq km was delineated.[13]

The China Ocean Mineral Resources Research and Development Association (COMRA) was eventually established in 1990. COMRA was defined as a state enterprise at that time.[14] The purpose of COMRA was to develop new mineral resources for China, enhance the development of high technology for deep seabed mining, and contribute to mankind's development and utilization of international seabed resources. The supreme authority of COMRA is the Council, which consists of 47 council members from 13 governmental departments and their subsidiaries. The implementing organ is the permanent council while the general office deals with daily affairs in accordance with the decisions of the Council and the Permanent Council.[15]

A milestone national conference was held in October 1999 regarding China's future strategy for deep seabed activities. The following consensus was reached through the conference: (a) to formulate a national strategy for the Area; (b) to adhere to the policy of being actively involved in activities in

13 The delineation followed some indexes: average nodule abundance of the area greater than 5 kg/m^2, nodule grade (content of copper, cobalt and nickel) greater than 1.8%, and the seabed slope less than 5^0. Wang Zhixiong, 'China and the Exploitation of Deep Seabed Polymetallic Nodules', *Marine Policy*, vol. 15, no. 2), 1991, p. 133.

14 Wang, *supra* note 13, p. 133.

15 See 'China Ocean Mineral Resources Research and Development Association', available at <http://www.soa.gov.cn/jigou/2/dayang.htm> accessed 4 September 2017.

the Area and to strengthen China's status in Area affairs and in the international arena; (c) to adopt the policy of 'continuously carrying out deep sea surveys, greatly developing deep sea technology, and establishing in due time a deep sea industry'; and (d) to single out three stages of development for the first half of the twenty-first century: resource prospecting and mining site application, studying and developing deep sea technology, and establishing a deep sea industry.[16]

COMRA is also directly engaged in international cooperation concerning deep seabed mining. In 1993 COMRA signed three agreements respectively: with the Interoceanmetal Joint Organization (IOM) concerning cooperative experiments on hydraulic lifting of polymetallic nodule ore pulp using pumped water; with the Nansen Institute of Norway concerning regional cooperation in deep seabed mining, strategic research, mining technology and other areas; and with the International Ocean Institute (IOI) concerning the organization of a training programme in the management of international seabed mining held in China in 1995.[17] In addition, COMRA was required as a pioneer investor to 'provide training at all levels for personnel designated by the Commission'.[18] Four trainees (from Algeria, Belarus, South Korea and Sudan) selected by the Commission arrived in China in mid-May 1994 to begin their training – COMRA fulfilled its training obligations on 25 April 1995.[19]

4 China and PrepCom

After the adoption of the UNCLOS, the PrepCom was established in December 1982 when 117 States and two other entities signed the Convention. This was in conformity with Resolution I contained in the Final Act of the UNCLOS III in that the UN Secretary-General should convene the Commission, and the

16 See 'Chronology of China's Research and Development of Ocean Mineral Resources, 1998–1999', available at <http://www.soa.gov.cn/dygz/98-99.htm> accessed 4 September 2017.

17 See 'Periodic Report on the Activities of the China Ocean Mineral Resources Research and Development Association (COMRA) in the Pioneer Area', ISBA/3/LTC/R.4, 24 June 1997, 4–6.

18 Resolution II, Para.12 (a)(ii).

19 See 'Training programmes under resolution II of the Third United Nations Conference on the Law of the Sea', ISBA/7/LTC/2, 25 June 2001, 2, 4 and 5; and 'Periodic Report of the China Ocean Mineral Resources Research and Development Association (COMRA) on Its Activities in the Pioneer Areas from 1 January to 31 December 1995', ISBA/4/LTC/R.5, 4 August 1999, at 3.

Commission should meet within 90 days thereafter, upon signature of or accession to the Convention by 50 States. In 1983 the PrepCom held its first session where a Plenary and four special commissions were established to implement the mandates entrusted by Resolution II. The most important result from the PrepCom meetings lie in two areas: the arrangement for the registration of pioneer investors based on Resolution II, and the adoption of the 1994 Agreement.

As referred to in Resolution II, pioneer investors include: (a) France, India, Japan and the former Soviet Union (now replaced by Russia) or their enterprises; (b) four entities whose components are from more than one of Belgium, Canada, former West Germany (now Germany), Italy, Japan, the Netherlands, the United Kingdom or the United States; and (c) any developing country or its enterprises.[20] Pioneer investors have the right to conduct pioneer activities such as undertakings, commitments of financial and other assets, investigations, findings, research, engineering, development and other activities relevant to the identification, discovery, and systematic analysis and evaluation of polymetallic nodules and to the determination of the technical and economic feasibility of exploitation. Pioneer investors can have a pioneer area of no more than 150,000 sq km allocated by the PrepCom for pioneer activities. Half of the allocated pioneer area should be relinquished to revert to the Area in due time in accordance with the schedule set forth by Resolution II.[21] However, there are a set of requirements for a country, or its entity to be registered as a pioneer investor: it should have signed the UNCLOS, identify an area of pioneer activities in its application, and pay a fee of US$250,000. The intended State had spent US$30 million for its pioneer activities before 1 January 1983 (1 January 1985 if the State was a developing country). Since no pioneer investor could be registered in respect of more than one pioneer area, the issue of overlapping claims among some countries needed to be resolved. To this end, the Arusha Understanding was reached in 1986, which laid down new procedures and a timetable for the resolution of overlapping claims.

Based on the Arusha Understanding, a prospective pioneer investor may allocate 52,300 sq km for itself and France, Japan and the Soviet Union together would give the ISA an area of 52,000 sq km. This procedure deviated from the original provision contained in Resolution II, but it was a compromise

20 1(a) of Resolution II.

21 The schedule is: (a) 20% is relinquished by the end of the third year from the date of the allocation; (b) additional 10% by the end of the fifth year from the date of the allocation; and (c) the final 20% after eight years from the date of the allocation or the date of the award of a production authorization, whichever is earlier.

to woo more pioneer investors. Since India's allocated area was in the Indian Ocean where no overlapping claims existed, India was the first to be registered through the PrepCom as a pioneer investor on 17 August 1987.[22] Encouraged, IFREMER/AFERNOD, DORD, and Yuzhmorgeologiya were registered at the end of that year. COMRA was registered on 5 March 1991, followed by IOM (21 August 1991) and the Government of the Republic of Korea (2 August 1994).

At the summer meeting of the PrepCom in 1991, China's obligation as a pioneer investor was discussed. Since the PrepCom treated the registration of India differently from other pioneer investors and imposed fewer obligations on it, China requested the same treatment on the grounds that both were developing countries. With much effort made by the Chinese delegation and repeated consultations at the meeting, the issue of China's obligation was finally resolved in a reasonable manner.

The 1994 Agreement and Part XI of the UNCLOS should be interpreted and applied together as a single instrument. In the event of any inconsistency, the provisions of the Agreement prevail.[23] The Agreement can override and is thus superior to the provisions of Part XI. This practice is unique in the sense that a treaty has been amended before its entry into force. The Agreement came into effect on 28 July 1996. It has changed, to some fundamental extent, the original deep seabed regime created by Part XI of the UNCLOS, though the principle of the common heritage of mankind continues to be maintained.[24] The major changes are reflected in the following eight areas:

(a) The burden of costs on States parties has been reduced;
(b) The privileges of the Enterprise have been cancelled so that it will become a normal enterprise;
(c) The new regulations have simplified the decision-making process;
(d) The special provisions on the Review Conference in Art. 155 (1)(3)(4) of the LOS Convention have been overridden by normal procedures of review and amendment;
(e) The obligation to transfer technology on a compulsory basis by contractors to the Enterprise has been cancelled;
(f) The production limit has been lifted;

---

22 The registration was described by the then UN Secretary-General Perez de Cuellar as 'one of the significant developments in the law of the sea since the adoption of the Convention'. Press Release SEA/876, 24 August 1987, at 1.

23 Art. 2 of the 1994 Agreement.

24 As provided in UNCLOS, Art. 311 (6), the principle of common heritage of mankind should not be amended.

(g) Developing land-based producer countries which are adversely affected by the production of minerals from the Area will be economically assisted by way of a fund to be established by the ISA;
(h) The procedure for financial terms of contracts has been simplified and the rates of payments reduced.

China actively participated in the consultations and it was one of the six proposing States (with Fiji, Germany, India, Indonesia and the United States) of the final draft Agreement. The Chinese delegate was satisfied with the result of the informal consultations and considered the 1994 Agreement a solution to the difficulties in implementing Part XI of the UNCLOS while maintaining the principle of common heritage of mankind and taking care of the countries that had or had not ratified the Convention. The Chinese delegate promised to apply *ad hoc* the Agreement from 16 November 1994 before China's official ratification.[25] It is noted that when China ratified the UNCLOS in 1996, it also ratified this Agreement.

5 China and the ISA

The ISA has been fully operational since June 1996. It has five subsidiary organs: the Assembly, the Council, the Legal and Technical Commission, the Finance Committee and the Secretariat. According to the UNCLOS, there should be a subsidiary organ under the Authority called "the Enterprise", which has yet to be established since no actual mineral exploitation activities have been conducted so far, and its function is currently performed temporarily by the Secretariat. Among the organs, the most important organ is the Council which, as a decision-making body of the ISA, has the power to lay down specific policies following the general policies by the Assembly. Due to its importance, the composition of the membership is very delicate: 36 members of the Authority are to be elected by the Assembly from five interested groups for the Council:

(a) Four members from those States Parties which, during the last five years for which statistics are available, have either consumed more than 2 per cent of total world consumption or have had net imports of more than 2 per cent of total world imports of the commodities produced from the

25 See Z. Lihai, *Studies on the Issues of the Law of the Sea*, Peking University Press, 1996, (in Chinese), pp. 186–187.

categories of minerals to be derived from the Area. The four members should include one State – the largest in terms of gross domestic product – from the Eastern European region.

(b) Four members from the eight States Parties which have the largest investments in preparation for and in the conduct of activities in the Area, either directly or through their nationals.

(c) Four members from States Parties which, on the basis of production in areas under their jurisdiction, are major net exporters of the categories of minerals to be derived from the Area, including at least two developing States whose exports of such minerals have a substantial bearing upon their economies.

(d) Six members from developing States Parties representing special interests. The special interests to be represented shall include those of States with large populations, States which are land-locked or geographically disadvantaged, island States, States which are major importers of the categories of minerals to be derived from the Area, States which are potential producers for such minerals, and less developed States.

(e) Eighteen members elected according to the principle of ensuring an equitable geographical distribution of seats in the Council as a whole, provided that each geographical region shall have at least one member elected under this subparagraph. For this purpose, the geographical regions should be Africa, Asia, Eastern European, Latin America and the Caribbean and Western Europe and Others.[26]

The first Council of the ISA was elected by the Assembly at the second session of the ISA in March 1996. China was elected as a member in Group B and reelected in 2000 for another four-year term from 2001 to 2004.[27] It is interesting to note that China was elected within Group B though China is eligible for candidacy for Groups D (developing countries) and E (regional representation). Since 2013, China has joined Group A – this reflects the change of China's status. In 1997, the Chinese delegation submitted to the ISA at its summer session the exploration plan of COMRA for the period 1998–2012, which was later approved by the Council of the Authority. The plan outlined the basic

26 The composition of the Council was originally provided for in UNCLOS, Art. 161(1), which has been replaced by a revised provision in Paragraph 15, Section 3 of the Annex to the 1994 Agreement.

27 See International Seabed Authority, *Handbook 2001*, May 2001, pp. 10–11.

framework for China's research and development activities relating to ocean polymetallic nodules in the Area.[28]

China participated in the deliberations of the mining code and stressed that the code should encourage qualified States or entities, particularly developing countries, to enter into the sphere of deep seabed activities.[29] During the first meeting of the sixth session of the ISA in March 2000 where the draft mining code was discussed, the Chinese delegation made several contributions to the progress of the negotiations. On the issue of confidentiality, the draft provisions favour the miners, containing legal liability and punishment for the disclosure of confidential materials. On the issue of environmental protection, the original wording of 'precautionary measures' was changed to 'precautionary approach' by the efforts of the Chinese delegation and others.[30] To China, precautionary measures or principles require miners to overspend on environmental equipment even before the occurrence of any evidence that the mining activities have caused environmental damage. The phrase 'best available technology' was also amended to 'best available technology to the miners' on the suggestion of China so that miners could avoid a potentially heavy financial burden caused by having to purchase the world's best technology. Finally, with the efforts of China, Russia and other countries, the procedure of reporting to the ISA was simplified. For example, the original separate requirements for marine environmental variations and marine monitoring were combined.[31] The Regulations on Prospecting and Exploration for Polymetallic Nodules in the Area were finally adopted in July 2000 and it was later updated and adopted on 25 July 2013.

Following the successful adoption of the Regulations on Prospecting and Exploration for Polymetallic Nodules in the Area, the ISA adopted the Regulations on Prospecting and Exploration for Polymetallic Sulphides in the Area (7 May 2010) and the Regulations on Prospecting and Exploration for Cobalt-Rich Crusts (27 July 2012). All the above three Regulations

28 'Periodic Report of the China Ocean Mineral Resources Research and Development Association (COMRA) on Its Activities in the Pioneer Areas from 1 January to 31 December 1997', ISBA/4/LTC/R.7, 4 August 1999, at 1.

29 Office of Policy Studies, Ministry of Foreign Affairs, PRC (ed.), *China Diplomacy 1999* (in Chinese), pp. 667–668.

30 As acknowledged, the scientific and legal implications of the application of the precautionary approach in the context of activities in the Area are extremely unclear and require considerable further study. See Lodge, *supra* note 5, at 288, no.106.

31 See 'Progress on the negotiation of the mining code at the first meeting of the six session of the ISA', available at <http://www.soa.gov.cn/intercop/ws15/nr6.htm> accessed 4 September 2017.

constitute the current 'mining code' for deep seabed mining. In March 2015, the Authority issued the following two consultation documents to members of the Authority and all stakeholders: Draft Framework for the Regulations of Exploitation Activities; and the Discussion Paper on the Financial Terms of Exploitation Contracts.[32] During the discussions, China expressed its views that while the draft made relatively comprehensive design and arrangements for marine environmental protection, it lacked similar design and arrangements for the development and utilization of seabed resources, particularly concerning the protection of operators. Therefore China urged that the draft should fully reflect the regulations and spirit of Part XI of the UNCLOS.[33]

In 2001, the ISA signed 15-year contracts with all existing pioneer investors except India for deep seabed explorations.[34] On 22 May 2001, the ISA officially signed the exploration contract with the COMRA in Beijing, formalizing the relationship between the two. The contract enables the COMRA to continue its exploration activities for polymetallic nodules in the northeast Pacific Ocean and gives the COMRA exclusive rights to explore the area allocated to it. The conclusion of the contract is a necessary step after a plan of work for exploration has been approved. As early as August 1997, the ISA approved the plans of work from all seven pioneer investors. China's approved plan would last 15 years.[35] Although the detailed terms and conditions of the Chinese contract are not yet available, the standard clauses annexed to the 2000 Regulations should be incorporated as required by the 2000 Regulations.[36] Furthermore, when the ISA concludes a contract with a State or entity, the conditions should be similar to and no less favourable than those agreed with any registered pioneer investor. Thus the principle of non-discrimination applies.

As of 31 January 2016, 23 contracts for exploration have entered into force (14 for polymetallic nodules, 5 for polymetallic sulphides and 4 for cobalt-rich ferromanganese crusts). The contracts sponsored by China are shown

---

32 'Ongoing Development of Regulations on Exploitation of Mineral Resources in the Area', available at <https://www.isa.org.jm/legal-instruments/ongoing-development-regulations-exploitation-mineral-resources-area> accessed 4 September 2017.

33 See Speech of Ma Xinmin, Deputy Dead of the Chinese Delegation to the 21st Session of the International Seabed Authority on the Agenda of 'Preparing the Regulations of Exploitation Activities', 22 July 2015, reprinted in *Chinese Yearbook of International Law*, Law Press, 2015, p. 742.

34 For details, see 'Deep Seabed Minerals Contractors', available at <https://www.isa.org.jm/deep-seabed-minerals-contractors> accessed 4 September 2017.

35 See Office of Diplomatic History Studies, Ministry of Foreign Affairs, PRC (ed.), *China's Diplomacy 1998* (in Chinese), p. 816.

36 Regulation 23 of the 2000 Regulations.

in Table 9.1 below. From the table below, we can see that China sponsored another Chinese enterprise in addition to COMRA for deep seabed mining. The Chinese contracts which expired in 2016 have now extended for another five years under ISA approval.

In recent years, China's involvement in ISA activities has intensified. On 3 March 2016, the ISA Legal and Technical Commission selected first-ranked and alternate candidates for the Global Sea Mineral Resources NV (GSR), China Ocean Mineral Resources Research and Development Association (COMRA) and Japan Oil, Gas and Metals National Corporation (JOGMEC) training programmes.

6 China's Legislation

In practice, deep sea mining within areas under national jurisdiction is regulated on the basis of legislation governing land-based mining in accordance with the Mineral Resources Law of the PRC, enacted in 1986 and amended in 1996.[37] This law provides a general legal framework for mining activities in China.

TABLE 9.1 *Contracts sponsored by China (as of 2016)*

1. Contracts for polymetallic nodules:
   COMRA, 22 May 2001, Clarion-Clipperton Fracture Zone, 21 May 2016;
   China Minmetals Corporation to be signed Clarion-Clipperton Fracture Zone (reserved area)
2. Contracts for polymetallic sulphides:
   COMRA, 18 November 2011, South-west Indian Ridge, 17 November 2026
3. Contracts for cobalt-rich ferromanganese crusts:
   COMRA, 29 April 2014, Western Pacific Ocean, 28 April 2029

SOURCE: PREPARED BY THE AUTHOR BASED ON ISBA/22/C/5, 'STATUS OF CONTRACTS FOR EXPLORATION IN THE AREA', 10 MAY 2016, AVAILABLE AT HTTPS://WWW.ISA.ORG.JM/SITES/DEFAULT/FILES/FILES/DOCUMENTS/ISBA-22C–5_1.PDF.

37 English text is available in Office of Policy, Law and Regulation, State Oceanic Administration (ed.), *Collection of the Sea Laws and Regulations of the People's Republic of China*, revised edition, Ocean Press, 1998, (in Chinese and English), pp. 233–252.

In 1994, the State Council passed the Rules for the Implementation of the Mineral Resources Law.[38] Art. 4 of the Regulations provides that the exploration and exploitation of the mineral resources within the territory of the PRC and other sea areas under its jurisdiction must comply with the Mineral Resources Law of the PRC and these Regulations.

China adopted a licensing-type system for mineral resources exploration and exploitation. Art. 13 and 16 of the Mineral Resources Law determine the competent agencies who are responsible for authorizing deep-sea mining: the department in charge of examination and approval of mineral reserves under the State Council or departments in charge of examination and approval of mineral reserves of provinces, autonomous regions and municipalities directly under the Central Government shall be responsible for the examination and approval of the prospecting reports to be used for mining construction design and shall, within the prescribed time limit, give official replies to the units that submitted the reports. Unless it is approved, a prospecting report may not be used as the basis for mining construction design.

Art. 16 provides that anyone who wishes to mine the following mineral resources shall be subject to examination and approval by the department in charge of geology and mineral resources under the State Council, which shall also issue a mining license:

(1) those within the mining areas embraced in State plans or within the mining areas which are of great value to the national economy;
(2) those outside the areas mentioned in the preceding sub-paragraph, and where the minable mineral reserves are at least of a large quantity;
(3) specified minerals of which protective mining is prescribed by the State;
(4) those in the territorial seas and other sea areas under China's jurisdiction; and
(5) other mineral resources as prescribed by the State Council.

The competent departments authorized by the State Council may conduct examination of and grant approval to mining of such specified minerals as oil, natural gas, radioactive minerals and issue mining licenses.

The mining of mineral resources that are not covered by the provisions of paragraphs 1 and 2 and the mineable reserves of which are of medium quantity shall be subject to examination and approval by the departments in charge of geology and mineral resources under the people's governments of

38 Available at <http://en.pkulaw.cn/display.aspx?cgid=9154&lib=law> accessed 4 September 2017.

provinces, autonomous regions and municipalities directly under the Central Government, which shall issue mining licenses. Measures for the administration of the mining of mineral resources not covered by the provisions of paragraphs 1, 2 and 3 shall be formulated by the standing committees of the people's congresses of provinces, autonomous regions and municipalities directly under the Central Government according to law. Where examination and approval are conducted and mining licenses are issued under the provisions of paragraph 3 and paragraph 4, the departments in charge of geology and mineral resources under the people's governments of provinces, autonomous regions and municipalities directly under the Central Government shall collect the cases and submit them to the department in charge of geology and mineral resources under the State Council for the record. The standards for large and medium quantities of mineral reserves shall be formulated by the department in charge of examination and approval of mineral reserves under the State Council.

Regarding maritime legislation, China has two basic laws: the 1992 Law on the Territorial Sea and Contiguous Zone and the 1998 Law on the Exclusive Economic Law and Continental Shelf.[39] Based on these two basic ocean laws, China established the maritime zones in accordance with the UNCLOS which China ratified on 15 May 1996. It is reported that China is preparing a comprehensive Ocean Law.

In terms of other related laws, there are the 1982 Marine Environment Protection Law, (amended 1999 and 2014), the 2001 Law on the Administration of the Use of Sea Areas, and the 2002 Law on Environmental Impact Assessment. The Law on the Administration of the Use of Sea Areas provides the regulation in details on the administration of the use of sea areas including the procedures on how to apply for use, the period of use and the approving process. Art. 25 of that Law provides the various lengths of period for the use of sea areas and the maximum period for salt production and mineral exploitation is 30 years.

Marine environmental protection is generally governed by the Marine Environment Protection Law (amended 1999 and 2014). Art. 46 provides general requirements on seabed mining: in building coastal construction projects, effective measures must be taken to protect wild animals and plants and their living environment as well as marine aquatic resources under State and local special protection. It is strictly prohibited to mine sand and gravel along the shore. In conducting coastal opencast mining or shore-based well drilling to

39 English versions of these two laws are appended to Zou Keyuan, *China's Marine Legal system and the Law of the Sea*, Nijhoff, 2005, pp. 338–345.

exploit seabed mineral resources, effective measures must be taken to prevent pollution to the marine environment. Currently, China's marine environmental law can be divided into two parts: legislation on environmental issues, and legislation on marine resources issues. Environmental legislation focuses on the prevention and management of marine environmental pollution, while resources legislation focuses on the exploration and exploitation of marine resources. China has established a relatively comprehensive legal system on marine environmental protection, which includes the Regulations on the Prevention and Control of Pollution Damages to the Marine Environment by Coastal Engineering Construction Projects, Regulations on the Prevention and Control of Pollution Damages to the Marine Environment by Offshore Engineering Construction Projects. The legal system of marine resources legislation is equally comprehensive, including Fisheries Law, Law on the Administration of the Use of Sea Areas, and Law on the Protection of Wildlife. Other legislations are the Provisions on the Administration of the Protection and Utilization of Uninhabitable Islands and the Provisions on the Administration of the Right to Use Sea Areas.

There are still defects in China's legal regime on ocean issues. For example, it focuses more on preventing and managing pollution than protecting marine resources, more on exploring and exploiting marine resources than the sustainable use of marine resources. Moreover, it lacks coordination from one law to another, and there are areas still not covered by the existing marine legal system.

In 1994, the Chinese government adopted *China Agenda 21* based on the spirit of the United Nations Conference on Environment and Development in 1992. The Agenda focuses on the sustainable development and protection of marine resources, rule of law for ocean activities and the improvement of the oceanic environment protection mechanism. Following *China Agenda 21*, the State Oceanic Administration adopted *China Ocean Agenda 21* which covers various areas of marine sustainable development including marine industries, coastal management, islands, marine science and technology, natural disaster prevention and mitigation, marine living resources, and marine environmental protection.

In 2012 and 2013, the State Council published two important documents: the National Marine Economy Development during the "12th Five-Year Plan" Period and the National Marine Programs Development during the "12th Five-Year Plan" Period. These aim to improve the marine legal system, including the promulgation of implementing regulations for the Law on the Administration of the Use of Sea Areas, the Marine Environment Protection Law, Mineral Resources Law, Island Protection Law, Fisheries Law, and Marine Traffic Safety

Law. The documents also aim to strengthen maritime law enforcement and its supervision, to strengthen the guidance for local marine legislation, to support reform and innovative activities in coastal areas, and to a build standard system of marine ecosystem and environmental protection.

7 New Legislation

In 2014, the draft Law on Exploring and Exploiting Resources in Deep Seabed Area, drafted by the NPC Environmental and Resources Protection Committee, had been included in the Legislative Plan of the 12th National People's Congress Standing Committee. During the first session of the 12th National People's Congress held in March 2013, 31 People's Representatives (equivalent to Members of Parliament) proposed a motion on the legislation of ocean resources exploration and exploitation. Delegates proposed in the motion that the deep ocean resources in the international seabed area have strategic significance to states, but China has no relevant domestic legislation, they advised to develop an ocean resources exploration and exploitation law, not only actively fulfilling China's treaty obligations, but also safeguarding national interests.

The Report of the 18th Congress of the Communist Party of China (CPC) states that 'We should enhance our capacity for exploiting marine resources, develop the marine economy, protect the marine ecological environment, resolutely safeguard China's maritime rights and interests, and build China into a maritime power'.[40] From the Report of the 18th CPC Congress and the reform program of the new administration, we can see that the Chinese government attaches great importance to building China into a maritime power and implementing a marine development strategy. Exploring and exploiting marine resources is the foundation of a maritime strategy and maritime economy. China has realised the urgency to adopt the deep seabed mining law, as most of the industrialized countries in the world have already enacted domestic legislation on deep seabed mining. There is also a tendency that whether the State Sponsor has domestic legislation in this field will become a condition for that State's future activities in the Area, which made China's enactment of the deep ocean exploration and development law face more pressing pressure. Hence China needed the legislation on deep ocean exploration and exploitation, which provides legal protection for the involvement of China in exploration and exploitation of mineral resources in the Area.

40 Available at <http://www.china-embassy.org/eng/zt/18th_CPC_National_Congress_Eng/t992917.htm> accessed 4 September 2017.

The Law on Exploring and Exploiting Resources in Deep Seabed Area of the PRC was officially adopted on 26 February 2016 and came into force on 1 May 2016. It contains 7 chapters and 29 articles.[41] Deep Seabed Area refers to the seabed, ocean floor and subsoil thereof, beyond the limits of national jurisdiction of China and other countries. The Law is designed to regulate exploration and exploitation activities, promote research on deep sea science, technology and resource investigations, protect the marine environment, promote the sustainable use of deep seabed resources, and safeguard the common interest of mankind.

According to the Law, any PRC citizen, legal person, or other organization, before applying to the ISA, shall apply to the competent authority of the State Council. The approved applicant can conduct exploration and exploitation only after it has signed such contract with ISA. Transfer or any substantial change should be subject to the approval of the competent authority of the State Council.

As for marine environmental protection, the new Law has a chapter containing three provisions. Accordingly, the contractor should, within reasonable and feasible limits, adopt necessary measures to prevent, reduce and control pollution or other harm to the marine environment arising from its activities in the area of exploration and exploitation (Art. 12). The contractor should determine environmental baselines, evaluate possible impact on the marine environment from development activities, prepare and implement environmental monitoring schemes (Art. 13). The contractor should take necessary measures to protect and preserve rare or vulnerable ecosystems, endangered species, and marine biodiversity (Art. 14).

It is noted that there are no specific rules regulating environment impact assessment (EIA) in deep sea mining. In the Chinese legal system, there is the Law on Environmental Impact Assessment, but it only applies to mining activities within China's jurisdiction. It is recalled that in 2009, the Chinese government issued the Regulations on Environmental Impact Assessment of Planning, and Art. 8 of the Regulations provides that 'An environmental impact assessment of planning shall include the analysis, forecast and assessment of the following items: 1. the possible overall impact on the ecological system of the relevant regions, basins and sea areas as a result of the execution of planning; 2. the possible long-term impact on the environment and human health as a result of the execution of planning; and 3. the relationship between the economic & social benefits and the environmental benefits, and

41 Available at <http://www.soa.gov.cn/zwgk/fwjgwywj/shfl/201602/t20160229_50172.html> accessed 4 September 2017.

the relationship between the present interests and the long-term interests from the execution of planning'.[42] Unfortunately, the new Law does not provide more detailed stipulations on EIA and its procedures.

The Law further provides that the State supports research on deep sea science and technology and resource investigations. The competent authority of the State Council should monitor and inspect the activities of the contractor. The Law also provides legal liability arising from deep seabed mining activities.

It is applauded that 'the promulgation of this law will help China better fulfil its responsibilities as a sponsoring State. This law has been filed with the International Seabed Authority for the record'.[43] However, the Law is more like an administrative rule. It provides details on how to apply and approve the mining activities. It does not provide the responsibilities of a sponsoring State, nor detailed rights of a contractor. Second, it is curious why it uses the term 'Deep Seabed Area' instead of 'International Seabed Area' as provided for in the UNCLOS. Since the Law is very general, there should be detailed regulations to be worked out for the implementation of the Law. The coordination of the Law and the relevant regulations adopted by the ISA in terms of implementation. There might be a potential conflict between the two.

8 Conclusion

China's deep seabed exploration began under the centrally planned economic system, but after two decades, China's economic system has shifted fundamentally to that of a market-oriented economy. It affects China's attitude and policy on deep seabed mining. As a potential producer of deep seabed minerals, China may have conflicts of interest with land-based producers. Such potential conflicts may be resolved under the UNCLOS or the WTO framework. As a rising big power, China's commitment to the international deep seabed regime may invite close and strong attention from the international community. With the rapid growth of its economy, China needs more mineral resources.

---

42 Available at <http://en.pkulaw.cn/display.aspx?cgid=120685&lib=law> accessed 4 September 2017.

43 Statement by H.E. Ambassador NIU Qingbao, Head of Delegation of the People's Republic of China, At the 22nd Session of the International Seabed Authority, July 2016, available at: <https://www.isa.org.jm/sites/default/files/files/documents/china-en.pdf> accessed 4 September 2017.

The new Law provides a legal basis for Chinese companies to engage in more activities in the Area, though more detailed implementing regulations should be worked out in the near future.

Bibliography

Lodge, M.W., 'International Seabed Authority's Regulation on Prospecting and Exploration for Polymetallic Nodules in the Area', *Journal of Energy & Natural Resources Law*, vol. 20, no. 3, 2002.

Wang, Zhixiong, 'China and the Exploitation of Deep Seabed Polymetallic Nodules', *Marine Policy*, vol. 15, no. 2, 1991.

Zhao, Lihai, *Studies on the Issues of the Law of the Sea*, Peking University Press, 1996 (in Chinese).

Zou, Keyuan, *China's Marine Legal system and the Law of the Sea*, Nijhoff, 2005.

CHAPTER 10

Multinational Corporations and International Environmental Liability: International Subjectivity and Universal Jurisdiction (Backs and Forths after Kiobel)

Pablo Ferrara

1 Introduction: Stepping Aside to Keep Moving

This article underlines the strong necessity to step aside from traditional 'state-centered' conception of international subjectivity of multinational corporations in favor of 'universal jurisdiction' regarding cases that involve international environmental damage liability.[1] National courts and international tribunals would openly sue against private entities with international weight performing dicey environmental activities such as deep seabed mining in the Area as well as other activities in the high seas (in particular offshore drilling, oil transportation, marine dumping, exploitation of marine living resources, marine biology research).[2]

In sight of the increasing volume of interest and submissions for mining in the Area – plus the additional outcome of a series of outstanding pollution disasters – it is a good time for a break and a deep thought on the necessity of a change in the jurisdictional approach.[3]

---

* Dr. Pablo Ferrara is legal technician at the Argentine Congress and Off Counsel at Estudio O'Farrell (Buenos Aires, Argentina).

1 See P.W. Birnie and A.E. Boyle, *International Law and the Environment*, Clarendon Press, 1992; L. Bergkamp, *Liability and Environment*, Kluwer Law International, 2001; M.-L. Larsson, *The Law of Environmental Damage*, Kluwer Law International, 1999; M.G. Faure, H. Lixin and S. Hongjun, *Maritime Pollution Liability and Policy*, Wolters Kluwer, 2010.

2 M. Karavias, *Corporate Obligations Under International Law*, Oxford University Press, 2014.

3 In 2005, the Council, on the recommendation of the Legal and Technical Commission, designated sectors W2 and E2 of the area under application (18,462 km^2 and 54,284 km^2 respectively, the total being 72,746 km^2) as the reserved area (Decision of the Council relating to a request for approval of a plan of work for exploration by the Federal Republic of Germany represented by the German Federal Institute for Geosciences and Natural Resources, ISBA/11/C/10, 42). The first seven reserved areas were contributed by the registered pioneer

 | DOI 10.1163/9789004360273_011

Regarding methodology, this is a *de lege ferenda* work – not a *de lege lata* discussion – based on ocean activities, existent national and international legal regimes and a conclusive argumentative basis. It is clear, then, that the aim of this work is to shed a hint of light upon a new line of useful legal environmental thought.

2 Mining in the Area: Single Coin, Several Sides

2.1 *Take Off*

On 1 November 1967, the Maltese Ambassador to the United Nations, Arvid Pardo, suggested to the United Nations General Assembly (UNGA) that the deep seabed and ocean floor and its common resources should be declared 'common heritage of mankind' (CHM).[4] Three years later, on 17 December 1970, the UNGA adopted the Declaration of Principles Governing the Seabed and Ocean Floor, and the Subsoil thereof, beyond the Limits of National Jurisdiction.[5] Ambassador Pardo's proposal was only the prelude to a large and conflictive series of negotiations in the following context of the 1982 United Nations Convention on the Law of the Sea (UNCLOS).[6] In any case, for some time already had the main mining corporations and their supporting states been doing feasibility research for deep seabed mining.[7] Interests were upon collision.[8] Public actors decided, then, to choose their plenipotentiaries and to leave the matter to negotiation. Nevertheless, eight years of debate and great expectations could not compensate the disappointment of UNCLOS' Part XI, as the text – including its references to the CHM regime – seemed somehow 'tendentious' both for mining corporations and their supporting States.[9] This circumstance, along with the retrieval of mining markets' demand, led offshore projects to stagnation and somnolence for twelve years throughout a full reordering of world geo-politics, global economy, and mineral market evolution.

---

investors at the time of the Preparatory Commission for the International Seabed Area and for the International Tribunal for the Law of the Sea: <https://www.isa.org.jm/contractors/reserved-areas> accessed June 2017.

4 UNGA GAOR 22nd Session (1 November 1967), p. 2.

5 UNGA Res 2749 (XXV) (12 December 1970).

6 UNGA GAOR 22nd Session (1 November 1967), p. 12.

7 A.M. Post, *Deepsea Mining and the Law of the Sea*, Martinus Nijhoff, 1983, p. 57.

8 Ibid., p. 116.

9 K. Maslar, *The Concept of the Common Heritage of Mankind in International Law*, Martinus Nijhoff, 1998, p. 206.

2.2 *Reappraisal: The 1994 Seabed Disputes Chamber Advisory Opinion*

On 17 August 1994, almost forty years after the first signs of interest, mining corporations and their neoliberal supporting governments decided it was a good time to move on.[10] After new negotiations and a new agreement, the UNCLOS mechanism was finally initiated and the Seabed Disputes Chamber (SDC) of the United Nations International Tribunal for the Law of the Sea (ITLOS) received a request for an Advisory Opinion (the Opinion) from the International Seabed Authority (ISA) regarding 'Responsibilities and obligations of States sponsoring persons and entities with respect to activities in the Area'.[11]

Regarding international subjectivity, the Opinion of the SDC is at first sight strongly State-centered and, possibly, rather 'conservative' in a doctrinarian sense due to the fact that it conveys the consciousness of the lack of 'materialization of the international subjectivity' to actors performing activities in the Area (i.e., corporations). In contrast, many modern literature scholars have discarded the traditionally overwhelmingly State-centered conception of the international legal framework. For them, 'the State is not considered a historical fact, but a legal status; individual freedom is not regarded best preserved inside the State, but by making it a concern for the international community as a whole; (...) factual developments do not have direct legal value, but have to be transformed into law through principled justification.'[12] In this line of thought, an updated contemporary legal order must outline an 'inclusive' framework of reference for international personality.[13]

As already indicated, modern literature scholars seem to have put aside the traditional international legal framework in exchange for an updated contemporary legal order, with the following milestones: (i) the openness of international law as a legal system from which no entity is *a priori* excluded; (ii) the inexistence of natural or primary as opposed to derived or secondary international persons.[14] In addition, according to the new international literature:

> [T]he personal scope of an international rule, with the exception of certain fundamental principles in which there is a presumed personality of individuals, is determined according to general rules of interpretation without starting from a particular presumption. In this process,

---

10 Ibid., p. 216.

11 *Responsibilities and Obligations of States Sponsoring Persons and Entities with Respect to Activities in the Area*, Advisory Opinion, ITLOS Reports 2011, p. 10 (Activities in the Area).

12 R. Portmann, *Legal Personality in International Law*, Oxford University Press, 2010, p. 282.

13 Ibid., p. 283.

14 Ibid., p. 269.

> recognition of international legal status may have considerable evidential value, but it is not in itself necessary or sufficient for constituting personality in international law; similar considerations apply with respect to effective actor status. It ensues that the sole consequence of being an international person is to be able to invoke international responsibility and to be held internationally responsible as far as applicable secondary rules exist. Crucially, it is not a consequence of personality to have the capacity to create international law.[15]

According to the abovementioned rule, it is possible to affirm that 'there is no sound basis for effective actor quality directly leading to international legal status',[16] and that the door is always open to new actors for the acquisition of international personality, provided they comply with the formal and individualistic requirements of the case:

> It is convenient to summarize the basic principles of personality in international law, which follow from a combination of the formal and individualistic conception:
>
> (1) International law is an open system. No entities are *a priori* excluded from it.
> (2) International personality follows from an international norm addressing an entity. It is thus acquired *a posteriori* through interpretation of specific norms. In case of treaty norms, normal rules of treaty interpretation apply. In principle, there is no *a priori* international person.
> (3) Special considerations apply when international crimes and fundamental human rights are concerned. In this context, related to the category of peremptory norms, there is a presumption that these basic principles directly apply to individual human beings, irrespective of them acting in official or private capacity. In this limited sense individuals are *a priori* international persons. Where analogy between individuals and legal persons of municipal law is not precluded on grounds of logic, the presumption arguably holds true for the latter. The legal status of States and other public entities in this context is established according to regular principles.

15 Ibid., p. 283.
16 Ibid., p. 269.

(4) Recognition is neither a necessary nor a sufficient condition for acquiring international personality. However, it has considerable evidential value when interpreting the personal scope of specific norms, especially those of customary nature. A similar function is exerted by effective actor status.

(5) The only direct consequence of possessing personality in international law is the capacity to invoke responsibility and to be held responsible for internationally wrongful acts (as far as applicable secondary rules exist). There is no further consequence of personality; in particular, there is no inherent capacity to create law.[17]

2.2.1 Multinational Corporations and Mining Activities

Under the framework outlined above, multinational corporations have been consolidating their international subjectivity during the last fifty years in several ways,[18] backed up by their substantial power in the international scenario (bargaining power, policy influence, political lobby, economic intervention, etc.).[19] On the latter, since the early beginning of the process of internationalization of corporations, mining companies have figured among the top multinationals.[20] The situation has not changed too much – except for the appearance of telecommunication and computer companies – and according to Price Waterhouse Coopers in 2013 there were six mining companies among the global top 100 capitalization multinational corporations (Table 10.1).[21] Now, if these companies' capitalization values were compared worldwide with the States' national GDP, they would fit between the 62nd and 51st strongest world economies (World Bank GDP report for 2012).[22] This provides an image of the potential power and interests at stake regarding mining corporations, their activities, associated States, and finally, relevant norms.

17 Ibid., p. 276.

18 P.T. Muchlinski, *Multinational Enterprises and the Law*, Oxford International Law Library, 2007, p. 18.

19 J.E. Nijman, *The Concept of International Legal Personality*, Asser Press, 2004, p. 354.

20 Cf. fn. 15.

21 Price Waterhouse Cooper's Global Top 100 Companies Infographics Report, <http://www.pwc.com/gx/en/audit-services/capital-market/publications/top100-market-capitalisation.jhtml> accessed June 2017.

22 World Bank list of national GDP per country, <http://data.worldbank.org/indicator/NY.GDP.MKTP.CD?order=wbapi_data_value_2012+wbapi_data_value+wbapi_data_value-last&sort=desc> accessed June 2017.

TABLE 10.1 *Mining companies among the Global Top 100 Capitalization Multinational Corporations*

Company Name	Nationality	Rank +/–	31 March 2013 Rank	Marke Cap $bn	31 March 2008 Rank	Market Cap $bn
BHP BILLITON	Australia	−7	28	171	21	175
RIO TINTO	United Kingdom	−35	65	92	30	155
VALE SA	Brazil	−41	67	90	26	159
BAYER	Germany	39	74	85	113	61
BASF	Germany	22	83	80	105	64
SABIC	Saudi-Arabia	−37	87	76	50	112

SOURCE: PWC INFOGRAPHICS REPORT.

2.2.2 Standardization

When discussion regarding deep-seabed mining reaches the level of law, national, transnational, and international interests collide. Of course, it is not easy to find a balance between the activity's inherent risks – both from the investors' lack of certainty regarding the mining spot and from the prospective environmental damage caused by the activity – and its extremely high potential profits. Nor is it easy to understand the importance to conceive a legal framework that allows a dangerous-but-highly-profitable activity and at the same time provides the broadest scope of insurance and securities for subjects not remotely involved in it (i.e. peoples from a far, land-locked, developing State), but who are supposedly entitled to benefits and protected in case of environmental damage by public international law. Even if not at the time of writing, this conundrum may become a harsh dilemma for the international community. And this dilemma is – of course – the *quid* of this essay.

3 The Area and the Common Heritage of Mankind

Whilst the *quid* of the present work is clear, it is interesting to analyze the possibilities of setting a point of departure in the analysis. In effect, there is one, and it is constituted by the concept of CHM.

The full scope of this concept is uncertain, but it is possible to ascertain some particular elements when applied to common space areas such as the seabed: (i) the areas cannot be appropriated; (ii) the use of the areas and their resources are governed and managed by an international authority; (iii) there is a regime of active and equitable sharing of benefits derived from the exploitation of resources; (iv) the peaceful use of the areas and resources concerned; and (v) the protection and preservation of the areas and resources for the benefit of mankind.[23] On the other hand, Professor Baslar affirms that 'the common heritage of mankind, as part of international resource management, is a functional concept rather than a territorial one,' and as such, 'it is not necessarily concerned with the ownership of the area where resources are found, but with the use of the resources for the benefit of mankind.'[24] Evidently, the *quid* of the CHM as a legal concept of international law is the management of resources; in particular, the fairness and justice connected with this management[25] and the subjects substantiating it.[26] On the latter, Professor Baslar reappraises:

> [T]he reappearance of mankind in international law constitutes a significant evidence of the demise of positivist school, because, starting from the Grotius' era until the second half of the 20th century, only States were deemed as the sole subjects of international relations having rights and duties. But at the dawn of the 21st century, it is believed that 'State sovereignty in international law is an idea whose time has come and gone.'[27] The resurgence of mankind pictures a shift from State-centered international law to anthropocentric international law, or from law of nations to law of mankind, or even from individualism to egalitarianism.[28]

Altogether, the anthropocentric character of the CHM concept may give way to the result that 'mankind' is no more than the steward of the earth charged with the protection of the earth and living organisms therein, that is, a *custodi in terram*.[29] Nevertheless, if the term 'mankind' is to be given an international

23 Cf. fn. 6.

24 Ibid., p. 49.

25 For a development on the approaches to resource management, see ibid. p. 57.

26 See H. Gould, *The Legacy of Punishment in International Law*, Palgrave Macmillan, 2010, p. 70.

27 R.B. Lillich, 'Sovereignty and Humanity: Can They Converge?', in A. Grahl-Madsen and J. Toman (eds.), *The Spirit of Uppsala: Proceedings of the Joint UNITAR-Uppsala University Seminar on International Law and Organization for a New World Order*, Walter de Gruyter, 1984, p. 105.

28 Cf. fn. 23, at p. 71.

29 Cf. fn. 6, at p. 65.

legal personality, the representation of all its components must be assured. Because of this difficulty, 'it is generally presumed that "the international community" represents mankind', and that 'the international system reflects more than ever before the growing importance of the idea of the international community: such a community is gaining a *sui generis* legal status distinct from its constituting nuclei (States).'[30]

4 The Common Heritage of Mankind and the Ocean

Professor Baslar states that '[w]hen Pardo introduced the common heritage on 1 November 1967, the protection of the marine environment against pollution of ships, radioactive waters or dumping was part of the common heritage of mankind.'[31] Professor Baslar is indeed a supporter of the idea that protection of the marine environment from the adverse effects of deep seabed mining has always been an element of CHM.[32] However, since the early 1980s, the concept of CHM has spread through the arguments of international environmental lawyers with a broader and independent sense. Thus international environmental law is claimed to have gained sufficient substance as to serve as a basis for a claim.

In 1969, the Legal Sub-Committee of the Seabed Committee was taken up on the discussion on the legal status of the area in question, providing the following comments:

> [T]he notion on the 'common heritage of mankind' would provide the basis for specific principles concerning the area; accordingly, all the rules and principles for activities in the sea-bed should be based on that notion. Its elements and consequences were: the notion of trust and trustees; indivisibility of the heritage; the regulation of the use of that heritage by the international community; the most appropriate equitable exploitation of this area to the developing countries; freedom of access and use by all States; and the principle of peaceful use.[33]

30 Ibid., p. 75.

31 Ibid., p. 235.

32 Ibid.

33 See UNGA 'Report of the Committee on the Peaceful Uses of the Sea-Bed and the Ocean Floor beyond the Limits of National Jurisdiction', UN GAOR 24th Session Supp. Nos. 22 and 22A, p. 14, paras 20–21.

In addition, the intention of Article 140 UNCLOS to differentiate between 'mankind,' 'States', and 'peoples' as different subjects is outlined when considering the economic and non-economic benefits of activities in the Area:

> Article 140
> Benefit of mankind
> 1. Activities in the Area shall, as specifically provided for in this Part, be carried out for the benefit of mankind as a whole, irrespective of the geographical location of States, whether coastal or land-locked, and *taking into particular consideration the interests and needs of developing States and of peoples who have not attained full independence or other self-governing status* recognized by the United Nations in accordance with General Assembly resolution 1514 (XV) and other relevant General Assembly resolutions.
> 2. The Authority shall *provide for the equitable sharing of financial and other economic benefits* derived from activities in the Area through any appropriate mechanism, on a non-discriminatory basis in accordance with Article 160, paragraph 2(f)(i).[34]

In giving effect to international rules and standards, Part XII UNCLOS explicitly acknowledges its framework character, and the existence of other global and regional sources in formulating rules and standards.[35] As Professor Redgwell put it, 'an echo of the consequences and obligations arising under other multilateral, regional or bilateral agreements is found in Article 282 of the LOSC',[36] which 'provides that a dispute regarding the interpretation or application of the LOSC shall be submitted to a procedure entailing binding decision thereunder, then that procedure will prevail over the procedures in Part XV of the LOSC, unless the parties to the agreement agree otherwise.'[37] In this regard, the ITLOS stated in the *MOX Plant Case* between Ireland and the United Kingdom that each and every environmental right and obligation,

---

34 United Nations Convention on the Law of the Sea (adopted 10 December 1982, entered into force 16 November 1994) 21 ILM 1261 (1982) (UNCLOS), emphasis added.

35 C. Redgwell, 'From Permission to Prohibition: The 1982 Convention on the Law of the Sea and Protection of the Marine Environment', in D. Freestone, R. Barnes, and D. Ong (eds), *The Law of the Sea: Progress and Prospects*, Oxford University Press, 2006, p. 181.

36 'LOSC' refers to UNCLOS.

37 Cf. fn. 30, at p. 183.

similar or identical, ran separately between UNCLOS and other legal multilateral or regional sources.[38] Such has always been the destiny of conventions on environmental matters in relation to UNCLOS: parallelism.[39] The evolution of international conventional environmental sea law has been quite substantial since 1982. Nevertheless, the basic set up in UNCLOS remains to be 'harmonization based on obligation of result (ie national standards shall be "no less effective" than global rules and standards, contained within and without the LOSC in preventing, reducing and controlling pollution of the marine environment by dumping).'[40]

Different international environmental obligations refer to the mining activities in the seabed. Their legal source can be similar, though independent. In any case, they are all linked to the CHM concept independently from seabed mining activities. Among the elements to support this assertion, there are six which are particularly relevant: (i) the difference in the nature of the international subject interested in the implementation of CHM;[41] (ii) the theoretical difference of the concept of 'justice' connected to the international obligation;[42] (iii) the difference of the jurisdictional location of the activities regulated by the international obligation;[43] (iv) the difference in share and burdens linked to the activity;[44] (v) the difference of scope and possibility to define clear limits;[45] and (vi) the presence or absence of manifestation of an element of stewardship and fiduciary responsibilities connected to the activities.[46]

5 The Opinion of the Seabed Disputes Chamber: Obligations and Responsibilities in Parallel

In relation to the sources of the obligations, the SDC was clear when it delivered its Opinion on the substantial burden of the States' primary obligations under UNCLOS:

38 *MOX Plant Case* (Ireland v. United Kingdom), Provisional Measures, ITLOS Reports 2001, 95.
39 Cf. fn. 32, at p. 184 et seq.
40 Ibid., p. 189.
41 Activities in the Area, para. 276.
42 Ibid.
43 Ibid.
44 Ibid.
45 Ibid.
46 Ibid.

> The role of the sponsoring State, as set out in the Convention, contributes to the realization of the common interest of all States in the proper application of the principle of the common heritage of mankind which requires faithful compliance with the obligations set out in Part XI. The common-interest role of the sponsoring State is further confirmed by its obligation, set out in Article 153, paragraph 4, of the Convention, to "assist" the Authority, which, as stated in Article 137, paragraph 2, of the Convention, acts on behalf of mankind.[47]

According to the SDC, the key provisions concerning the obligations of the sponsoring States are defined by: Article 139 (1); Article 153 (4) (especially the last sentence); and Annex III, Article 4 (4) UNCLOS.[48] Nevertheless, the definition of the term 'international community' is quite demanding a task, and attempting to interpret 'mankind' as the 'international community of "States"' seems to be rather narrow, stretch, and – altogether – misleading.

Regarding the actors in the Area, the SDC stated that pursuant to Annex III, Article 4 (3) UNCLOS, and confirmed in Regulation 11, Paragraph 2, of the Nodules Regulations and of the Sulphides Regulations, the SDC made the following statements regarding actors (i.e. domestic law entities) performing activities in the Area: (i) the connection between States Parties and domestic law entities required by UNCLOS is twofold, namely, that of nationality and that of effective control; (ii) all contractors and applicants for contracts must secure and maintain the sponsorship of the State or States of which they are nationals; and (iii) if another State or its nationals exercises effective control, the sponsorship of that State is also necessary.[49]

In relation to the act of sponsorship, the SDC stated that no provision of UNCLOS imposes an obligation on a State Party to sponsor an entity that holds its nationality or is controlled by it or by its nationals. The SDC clearly added:

> As the Convention does not consider the links of nationality and effective control sufficient to obtain the result that the contractor conforms with the Convention and related instruments, it requires a specific act emanating from the will of the State or States of nationality and of effective control. Such act consists in the decision to sponsor.[50]

47 Ibid., para. 76.
48 Ibid., para. 99.
49 Ibid., para. 77.
50 Ibid., para. 78.

Now, referring to the activities, the SDC opined in the first place, that 'the expression "activities in the Area" in the context of both exploration and exploitation includes, first of all, the recovery of minerals from the seabed and their lifting to the water surface'.[51] In addition, 'the transporting, processing and marketing of minerals recovered from the Area. Consequently, the latter activities are not included in the notion of "activities in the Area" referred to in Annex IV, Article 1, paragraph 1, of 1982 UNCLOS',[52] however 'transportation within that part of the high seas, when directly connected with extraction and lifting, should be included in activities in the Area',[53] due to the fact that 'inclusion of transportation to points on land could create an unnecessary conflict with provisions of 1982 UNCLOS such as those that concern navigation on the high seas'.[54] Finally, regarding exploration, though excluded from the UNCLOS definition of 'activities in the Area', the SDC considered it 'appropriate to observe that some aspects of the present Advisory Opinion may also apply to prospecting'.[55] In effect, the SDC noted in its Opinion that:

> [T]he Regulations issued to date by the Authority deal only with prospecting and exploration. Considering that the potential for damage, particularly to the marine environment, may increase during the exploitation phase, it is to be expected that member States of the Authority will further deal with the issue of liability in future regulations on exploitation.[56]

Evidently, the abovementioned arguments cause a collision between the obligations of the States *vis à vis* the obligations of the contractors. The SDC made clear that its primary addressees were States, either sponsoring the activities or corporately linked to the private entity performing them;[57] only

51 Ibid., para. 94.
52 Ibid., para. 84.
53 Ibid., para. 96.
54 Ibid., para. 96.
55 Ibid., para. 98.
56 Ibid., para. 168.
57 Ibid., para. 158. Regarding State labelling, the SDC stated that: 'Equality of treatment between developing and developed sponsoring States is consistent with the need to prevent commercial enterprises based in developed States from setting up companies in developing States, acquiring their nationality and obtaining their sponsorship in the hope of being subjected to less burdensome regulations and controls. The spread of sponsoring States "of convenience" would jeopardize uniform application of the highest standards of protection of the marine environment, the safe development of activities in the Area and protection of the common heritage of mankind'. See Activities in the Area, para. 159.

in the background did the SDC directly refer that the entities were responsible for deep seabed mining '"in conformity" or in "compliance" with the rules to which they refer'.[58] The Opinion is strongly focused on finding and holding the nexus between private actors and States in order to lead burdens and liabilities from the former to the latter:

> While it is not considered reasonable to make a State liable for each and every violation committed by persons under its jurisdiction, it is equally not considered satisfactory to rely on mere application of the principle that the conduct of private persons or entities is not attributable to the State under international law (see ILC Articles on State Responsibility, Commentary to Article 8, paragraph 1).[59]
>
> Under international law, the acts of private entities are not directly attributable to States except where the entity in question is empowered to act as a State organ (Article 5 of the ILC Articles on State Responsibility) or where its conduct is acknowledged and adopted by a State as its own (Article 11 of the ILC Articles on State Responsibility).[60]
>
> [T]he rules of the Convention concerning activities in the Area, although being treaty law and thus binding only on the subjects of international law that have accepted them, become effective for sponsored contractors which find their legal basis in domestic law. This mechanism consists in the creation of obligations which States Parties must fulfil by exercising their power over entities of their nationality and under their control.[61]

In respect to the content of the obligation, the SDC differentiated between two categories: (i) 'indirect obligations' of conduct or 'due diligence' obligations,[62]

---

58 These rules are referred to as 'this Part' (Part XI) in: (i) UNCLOS 1982 art 139, as 'the relevant provisions of this Part and the Annexes relating thereto, and the rules, regulations and procedures of the Authority, and the plans of work approved in accordance with paragraph 3'; (ii) UNCLOS 1982 art 153(4); and (iii) UNCLOS 1982 Annex III art 4(4) ('the terms of its contract and its obligations under this Convention'). See *Responsibilities and Obligations of States Sponsoring Persons and Entities with Respect to Activities in the Area*, Advisory Opinion, ITLOS Reports 2011, paras 103–104.

59 Ibid., para. 112.

60 Ibid., para. 182.

61 Ibid., para. 108.

62 In *Pulp Mills on the River Uruguay*, (Argentina v. Uruguay), Judgment, ICJ Reports 2010, p. 14, the ICJ illustrated the meaning of a specific treaty obligation that it had qualified as 'an obligation to act with due diligence' as follows:

and (ii) 'direct obligations'.[63] With respect to the first category, the SDC considered that it was not called upon to render specific advice as to the necessary and appropriate measures that the sponsoring State must take in order to fulfill its responsibilities under UNCLOS, as '[j]udicial bodies may not perform functions that are not in keeping with their judicial character'.[64] Nevertheless, it laid the foundation to understand the underlying law. In this sense, according to the SDC, the first category must not be understood as an obligation to achieve, but rather as 'an obligation to deploy adequate means, to exercise best possible efforts, to do the utmost'[65] under the following conditions: (i) sponsoring States' 'responsibility to ensure' applies 'within their legal systems',[66] taking into account, among other things, the particular characteristics of that system;[67] (ii) the measures taken by the States must be 'reasonably appropriate';[68] and (iii) the standard of due diligence has to be more severe for the riskier activities.[69] The second category is formed particularly by: '(i) the obligation to assist ISA in the exercise of control over activities in the Area; (ii) the obligation to apply a precautionary approach; (iii) the obligation to apply best environmental practices; (iv) the obligation to take measures to ensure the provision of guarantees in the event of an emergency order by the Authority for protection of the marine environment; (v) *the obligation to ensure the availability of recourse for compensation in respect of damage caused by pollution*; and (vi) the obligation to conduct environmental impact assessments'.[70]

"It is an obligation which entails not only the adoption of appropriate rules and measures, but also a certain level of vigilance in their enforcement and the exercise of administrative control applicable to public and private operators, such as the monitoring of activities undertaken by such operators (para. 197)".
Similar indications were given by the International Law Commission in its Commentary to Article 3 of its Articles on Prevention of Transboundary Harm from Hazardous Activities (2001).

63 Ibid., para. 121.

64 Ibid., para. 227.

65 Ibid., para. 110.

66 Ibid., para. 117.

67 Ibid., para. 229.

68 Ibid., para. 120.

69 Ibid., para. 117.

70 Ibid., para. 122, emphasis added.

6 Liability

According to the SDC, the framework for liability is set up by Article 139 (2) and Article 304 UNCLOS, which refer, respectively, to the 'rules of international law' as well as to 'the application of existing rules and the development of further rules regarding responsibility and liability under international law'. The SDC stated that account has to be taken of such rules under customary law, especially in light of the ILC Articles on State Responsibility.[71]

In the first place, liability arises from the failure of the sponsoring State to carry out its own responsibilities. The basic, general condition is that the sponsoring State is not liable for the failure of the sponsored contractor to meet its obligations.[72] There is, however, a link between the liability of the sponsoring State and the failure of the sponsored contractor to comply with its obligations, thereby causing damage.[73] Precisely, regarding the relationship between the contractor's liability and that of the sponsoring State, attention may be drawn to Annex III, Article 22 UNCLOS, which reads as follows:

> The contractor shall have responsibility or liability for any damage arising out of wrongful acts in the conduct of its operations, account being taken of contributory acts or omissions by the Authority. Similarly, the Authority shall have responsibility or liability for any damage arising out of wrongful acts in the exercise of its powers and functions including violations under Article 168, paragraph 2, account being taken of contributory acts or omissions by the contractor. Liability in every case shall be for the actual amount of damage.[74]

Consequently, there is only one point of connection, namely, that the liability of the sponsoring State depends upon the damage resulting from activities or omissions of the sponsored contractor.[75] Nevertheless, according to the SDC, this is merely a trigger mechanism, and such damage is not automatically

71 UNCLOS Art. 139(2) and the related provisions referred to above, prescribe or refer to different sources of liability, namely, rules concerning the liability of States Parties (UNCLOS Art. 139(2) first sentence), rules concerning sponsoring State liability (UNCLOS Art. 139(2) second sentence), and rules concerning the liability of the contractor and the Authority (referred to in UNCLOS Annex III art 22). See Activities in the Area, para. 171.

72 Ibid., para. 172.

73 Ibid., para. 173.

74 Ibid., para. 199.

75 Ibid., para. 181.

attributable to the sponsoring State: 'If the contractor has paid the actual amount of damage, as required under Annex III, Article 22, of 1982 UNCLOS in the view of the Chamber, there is no room for reparation by the sponsoring State'.[76] The situation becomes even more complex if the contractor has not fully covered the damage. It was pointed out in the Opinion proceedings that 'a gap in liability may occur if, notwithstanding the fact that the sponsoring State has taken all necessary and appropriate measures, the sponsored contractor has caused damage and is unable to meet its liability in full'.[77] It was further pointed out that 'a gap in liability may also occur if the sponsoring State failed to meet its obligations but that failure is not causally linked to the damage'.[78] In the Opinion's written and oral statements, States Parties to UNCLOS expressed different views on this issue; some argued that the sponsoring State had a residual liability, that is, the liability to cover the damage not covered by the sponsored contractor although the conditions for a liability of the sponsoring State under Article 139 (2) UNCLOS were not met; other States Parties took the opposite position. In the view of the SDC, the liability regime established by Article 139 UNCLOS and related instruments leaves no room for residual liability: the liability of the sponsoring State and the liability of the sponsored contractor exist *in parallel*.[79]

In terms of conditions, Article 139 (2) UNCLOS clearly establishes two elements for liability to arise: (i) the failure by act or omission of the sponsoring State to carry out its responsibilities; and (ii) the occurrence of damage.[80] Neither UNCLOS nor the relevant Regulations specify what constitutes compensable damage, or which subjects may be entitled to claim compensation.[81] The SDC determined, then, that '[i]t may be envisaged that the damage in question would include damage to the Area and its resources constituting the common heritage of mankind, and damage to the marine environment'.[82] Regarding claimants, the SDC asserted that '[s]ubjects entitled to claim compensation may include the Authority, entities engaged in deep seabed mining, other users of the sea, and coastal States'.[83]

---

76 Ibid., para. 202.

77 Ibid., para. 203.

78 Ibid.

79 Ibid., para. 204.

80 Ibid., paras. 176 and 178.

81 See Regulation 30 of the Nodules Regulations and Regulation 32 of the Sulphides Regulations.

82 Ibid., para. 179.

83 Ibid., para. 179.

With regard to the standard of liability, it was argued in the proceedings that the sponsoring State has strict liability, i.e., liability without fault;[84] in the event of multiple sponsorship, liability is joint unless otherwise provided in the Regulations issued by the Authority.[85] Compensation shall be for the actual amount of damage, that is, full compensation or *restitutio in integrum*.[86] Finally, it is essential to consider that the nature of the obligation determines the scope of the liability.[87]

The availability of recourse for compensation was established by the SDC as follows: '[I]t is inherent in the "due diligence" obligation of the sponsoring State to ensure that the obligations of a sponsored contractor are made enforceable'.[88] In this sense, Article 235 (2) UNCLOS reads:

> States shall ensure that recourse is available in accordance with their legal systems for prompt and adequate compensation or other relief in respect of damage caused by pollution of the marine environment by natural or juridical persons under their jurisdiction.

Support for the enforcement of the contractor's obligations under the domestic law of the sponsoring State is an essential requirement in a number of national jurisdictions.[89] UNCLOS specifies in various provisions the issues that should be covered by the sponsoring State's laws and regulations. In particular, Article 39 of the Statute dealing with enforcement of decisions of the SDC provides:

---

84 Ibid., para. 189.

85 Ibid., para. 192.

86 Ibid., paras. 193 et seq. This conclusion was first reached by the Permanent Court of International Justice in *Case Concerning the Factory at Chorzów* (Germany v Poland) (Merits) PCIJ Rep Series A No 17. The International Law Commission further reiterated this obligation. According to ILC Articles on State Responsibility Art. 31(1): 'The responsible State is under an obligation to make full reparation for the injury caused by the internationally wrongful act'. The Chamber notes in this context that treaties on specific topics, such as nuclear energy or oil pollution, provide for limitations on liability together with strict liability. The Chamber wishes to refer to ILC Articles on State Responsibility Art. 34. It reads: 'Full reparation for the injury caused by the internationally wrongful act shall take the form of restitution, compensation and satisfaction, either singly or in combination in accordance with the provisions of this chapter'.

87 Activities in the Area, para. 177.

88 Ibid., para. 239.

89 Ibid., para. 218.

> The decisions of the Chamber shall be enforceable in the territories of the States Parties in the same manner as judgments or orders of the highest court of the State Party in whose territory the enforcement is sought.[90]

Finally, reference must be made to Annex III, Article 21 (2) UNCLOS, which provides:

> Any final decision rendered by a court or tribunal having jurisdiction under this Convention relating to the rights and obligations of the Authority and of the contractor shall be enforceable in the territory of each State Party.

In a number of national jurisdictions, these provisions may require specific legislation for implementation. In any case, they certainly are mandatory, but against whom: Who should be sued to ensure compensation? To this question, the SDC answered that these obligations apply both 'to the sponsoring State' as well 'as the State with jurisdiction over the persons that caused the damage'.[91] Nevertheless, no reference was made to the private entities performing the activities or to any possibility of co-demanding them in case of collusion to avoid compliance of obligations of 'due diligence.'

7 Actio Potestas: Where Parallels Meet

It is possible to understand that the necessity for a rather impartial 'active hand' taking care of the CHM refers at a certain point to the 'Public Trust Doctrine',[92] and consequently to the opportunity to recognize a 'trustee' dealing with the specific interests of 'mankind'. This has many times been the point of departure for the discussion about the creation of a CHM authority as well as about the creation of reserve funds.[93] In a sense, it seems that the international subject has a 'collective soul', but unfortunately, no 'recognizable body'. It would

---

90 Ibid., para. 235.

91 'By requiring the sponsoring State to establish procedures, and, if necessary, substantive rules governing claims for damages before its domestic courts, this provision serves the purpose of ensuring that the sponsored contractor meets its obligation under Annex III, Article 22, of the Convention to provide reparation for damages caused by wrongful acts committed in the course of its activities in the Area'. Activities in the Area, para. 140.

92 Cf. fn. 6, at p. 65.

93 Cf. fn. 6, at pp. 374 et seq.

therefore be essential for the completion of this legal institution to provide the 'soul' with a proper 'body'. An answer to this might be the installment of an adequate *actio potestas*.[94]

8 Traditional Disputes and Contemporary Bargain

The distinction between 'State' and 'nation' (and the parallel differentiation between both of those terms and 'sovereign') is part of what caused 'international punishment' to fall out of general favor.[95] Nations, as congeries of natural persons, were regarded as bound by natural law; States, as artificial persons distinct from the natural persons they personated, were progressively harder to portray as bound by natural law.[96]

There were three main components to classical 'international punishment': (i) absolutely and universally binding nonconsensual rules; (ii) universal standing to punish those who violate the rules without regard to injury or interest; (iii) and an understanding of sovereignty without implications of inviolability or the sort of immunity from judgment that comes from juridical equality.[97] The first pillar of international punishment were absolute rules of universal scope; the second – logically dependent upon the first – was the universal standing to enforce those rules.

Modern international law has conceptualized obligations as bilateral in character even when the instruments declaring them were multilateral in scope. International law does not generally understand obligations to be owed to the community of States at large. Standing to seek redress for the violation of a rule was (and remains) overwhelmingly limited to parties directly and materially injured by the violation. There was nothing necessary about this: bilateral agreements were simply more common historically, and bilateral obligation was the cognitive framework within which early multilateral agreements were formulated, and this conceptualization has been utilized ever since.

The last decades of the 20th century saw the enunciation – if not clear-cut application – of a new manifestation of the second pillar. Since the 1970s,

94 'Ideally, a non-state representation of future generations is needed should the common heritage of mankind be effectively implemented. However, this is a big challenge for present international law'. n 6, at p. 78.

95 Cf. fn. 23, at p. 66.

96 Cf. fn. 6, at p. 7.

97 Ibid., p. 8.

international law recognized the existence of obligations *erga omnes*, that is, obligations that were owed generally to the international community as a whole. The basic *postulatio* regarding the violation of these obligations, since they were owed to the international community generally, was that it gave rise to general standing to enforce them, namely, to the *actio popularis*.

Based on a practice from Roman law, the *actio popularis* is an action brought against a malfeasant not on the personal behalf of the party bringing the action, but on behalf of the community. In international law, it is one State making a legal claim against another not for an injury it suffered by the conduct of the latter, but for the latter's mere violation of an obligation owed *erga omnes*. This practice is well known in many domestic legal settings, but it has been greeted simultaneously with trepidation and scepticism in international law.

The emphasis with obligations *erga omnes* is not on violations of natural law punishable by any and all, but rather on violations of positive law, voluntarily undertaken obligations owed to all, legally actionable by any. In principle, every State has a legal interest in the observance of obligations *erga omnes*; thus, it has been asserted that any State may take legal action to ensure compliance with an obligation *erga omnes*. In this sense, '[p]ositivists, on the one side, apologize that the traditional sources of international law do not give due credibility to the common heritage of mankind as a legal principle. Whereas, on the other side, non-positivits claim that the common heritage of mankind is an established legal principle of international law, acquiring its legal normativity from other sources such as general principles of law and even from peremptory norms'.[98] In order not to oscillate between 'apology' and 'utopia', Professor Baslar suggests that 'the view that the difficulty of bringing the common heritage of mankind into the corpus of international law may be surmounted by having recourse to the Grotian or eclectic philosophy'.[99] In the end, 'the common heritage of mankind as a distinctive illustration of idealism relies on different sources like inter alia, resolutions and declarations of international organizations, quasi-legislative acts of the United Nations and *jus cogens*'.[100] From the current perspective, the goal is to internationalize the CHM concept as a general principle of law inherent to every human being, substantiating the naissance of a 'human right' per se:

98 Ibid., p. 367.

99 Ibid., p. 368.

100 Ibid., p. 359.

> [I]f the concept of common heritage of mankind was, first of all, transposed into the human rights framework, it might be possible to justify human rights law on a non-conventional and non-consensual basis by rejecting customary law and opting rather for general principles of law. Accepting the common heritage of mankind as a human right would also enable us to justify the right to intervention in internal affairs on behalf of mankind and the right to *locus standi* or *actio popularis* in the internal affairs of sovereign states at a time when humanity's heritages are misused.[101]

It is precisely at this point that the question arises: What about the possibility of an *actio popularis* for environmental damages with universal *locus standi?*

9 International Tendency: The Opinion of the Seabed Disputes Chamber

As previously discussed, when analyzing breaches of 'due diligence' obligations, no mention was made by the SDC of the possibility of bringing private entities to court in case of environmental sea damage. In addition, the SDC made reference only indirectly to the existent substantive obligations of public international environmental law – when redirecting them to Part XI UNCLOS, according to Article 209.[102] In any case, the SDC opened up the possibility of an *actio popularis* for the Authority:

> No provision of the Convention can be read as explicitly entitling the Authority to make such a claim. It may, however, be argued that such entitlement is implicit in Article 137, paragraph 2, of the Convention, which states that the Authority shall act 'on behalf' of mankind. Each State Party may also be entitled to claim compensation in light of the *erga omnes* character of the obligations relating to preservation of the environment of the high seas and in the Area.[103]

In support of this view, the SDC made reference to Article 48 of the ILC Articles on State Responsibility, which provides:

---

101 Ibid., p. 369.

102 Activities in the Area, para. 241.

103 Ibid., para. 208.

> Any State other than an injured State is entitled to invoke the responsibility of another State (...) if: (a) the obligation breached is owed to a group of States including that State, and is established for the protection of a collective interest of the group; or (b) the obligation breached is owed to the international community as a whole.[104]

Now, it is important to consider that even if the SDC recognized the existence of the environmental obligations *erga omnes* and the existence of an *actio popularis* to implement them, unfortunately it did not recognize the possibility to bring private entities to court.

9.1 *The Liability Gap*

When analyzing the liability *ménage*, the SDC found itself dealing with what it defined as 'a gap in the liability regime established in Part XI of 1982 UNCLOS and related instruments' regarding the impact of international law on the deep seabed liability regime. In order to find a solution to that gap, the SDC stated that the provisions in Articles 139 (2), first sentence, and Article 304 UNCLOS are '"without prejudice" to the rules of international law'.[105] However, the SDC noted that 'if the sponsoring State has not failed to meet its obligations, there is no room for its liability under Article 139, paragraph 2 UNCLOS even if activities of the sponsored contractor have resulted in damage. A gap in liability which might occur in such a situation cannot be closed by having recourse to liability of the sponsoring State under customary international law'.[106] Then the SDC affirmed that although it was aware of the efforts made by the ILC to address the issue of damages resulting from acts not prohibited under international law, such efforts have not yet resulted in provisions entailing State liability for lawful acts. The SDC understood, therefore, that any damage produced under those circumstances was uncovered, drawing the attention of the Authority to the option of establishing a trust fund to cover such damages. Thus, upon unfortunate conceive of the SDC, private entities are to be kept out of the possibilities of a court claim in view of current international law.

104 Ibid., para. 180.

105 Ibid., para. 209.

106 Ibid.

10 National Tendency Extremes: United States of America and Spain

A brief reference shall hereby be made to the United States and Spain in order to provide an idea of how the elements of the current debate are treated in national jurisdictions.

10.1 *The United States Alien Tort Claims Act: Ups and Downs in the Stairway to Preclusion*

As Georg Nolte recalls, 'civil judgments which reply on the principle of universal jurisdiction in cases of asserted violations of international law have mainly occurred in the United States (...) based on the Alien Tort Claims Act of 1789'.[107]

The United States Alien Tort Claims Act (the Act) establishes the jurisdiction of the United States federal courts for 'all cases where an alien sues for a tort only in violation of the law of nations or a treaty with the United States'.[108] On a case-by-case basis, the Act has traditionally been applied in courts in cases dealing with 'universally recognized principles of international law' or accepted norms of the international law of human rights,[109] upholding protection against either people or incorporated entities.[110] Unfortunately, the violation of international environmental law has not fallen under the Act yet.[111]

The Act itself does not describe a special category of law, but simply speaks of 'violation of the law of nations'. However, 'American courts have always insisted on another requirement: that the relevant norm of the "law of nations" (should) be clear and specific'. In its most recent case, the US Supreme Court accepted the existent 'specific, universal, and obligatory' requirement standard.[112] Pursuant to this judgment, some authors understand that 'for the American judiciary it is not the substance, but the form of an international

107 G. Nolte, 'Universal Jurisdiction in the Area of Private Law – The Alien Claims Act', in C. Tomuschat and J.M. Thouvenin (eds), *The Fundamental Rules of the International Legal Order*, Nijhoff, 2006, p. 373.

108 Alien Tort Statute 28 USC 1350.

109 J.-F. Flauss, 'Competence civile universelle et droit international général', in C. Tomuschat and J.M. Thouvenin (eds), *The Fundamental Rules of the International Legal Order*, Nijhoff, 2006, p. 406.

110 See *Filártiga v Peña-Irala*, 630 F.2d 876 (2nd Cir. 1980), *Kadic v Karadzic* (70 F.3d 232; 2nd Cir. 1995); *Doe v Unocal Corp* (248 F.3d 915; 9th Cir. 2002), *Ken Wiwa v Royal Dutch Shell Petroleum Co. et al.* (226 F.3d 440; 2nd Cir. 2000).

111 See *Amlon Metals Inc. v FMC Corp.* 775 F. Supp. 668 (SDNY 1991) 670.

112 See *Sosa v Alvarez Machain*, 542 US 692 (2004).

legal norm that is relevant; not its importance but its specificity; not its moral obviousness, but its visible international acceptance in practice'.[113]

Regarding functioning and reach of US tribunal's hand under international law, it remains a preliminary question ruled by the 'Charming Betsy Rule' of interpretation of domestic law in conformity with international law.[114] On the other hand, the European Commission 'has underlined the necessity of recognizing the primary jurisdiction of local courts which are better suited to clarify and adjudge the issue'.[115] However, this could lead to unfair decisions in cases where corporations are to be sued and which prefer to proceed in their native courts, or else in a symbolic tool of political hegemony[116] through the exercise of extraterritoriality and privatization of the international jurisdiction of human rights.[117] Thus, a fair boundary to jurisdiction must be tailored, enabling the principle of 'reasonableness' through the presence of a foreign defender and the consideration of the doctrine of the *forum non conveniens.*[118]

Special mention deserves the latest Supreme Court leading case for the Act, *Esther Kiobel, Individually and on behalf of her late husband, Dr. Barinem Kiobel, et al. v.Royal Dutch Petroleum Co. et al.*, delivered on April 17, 2013 by Justice Robertson.[119] The Court defined clearly that '[t]he question here is not

113 Cf. fn. 102, at p. 376.

114 Ibid., p. 377. Also, Note 14 of the Brief of the Governments of the Commonwealth of Australia, the Swiss Confederation, and the United Kingdom of Great Britain and Northern Ireland as Amicus Curiae in *Sosa v Alvarez Machain.*

115 Ibid., p. 379.

116 J.-F. Flauss, 'Competence civile universelle et droit international général', in C. Tomuschat and J.M. Thouvenin (eds), *The Fundamental Rules of the International Legal Order*, Nijhoff, 2006, 389.

117 Ibid., p. 391.

118 Ibid., p. 395.

119 Petitioners were residents of Ogoniland, an area of 250 square miles located in the Niger delta area of Nigeria and populated by roughly half a million people. When the complaint was filed, respondents Royal Dutch Petroleum Company and Shell Transport and Trading Company plc, were holding companies incorporated in the Netherlands and England, respectively. Their joint subsidiary, respondent Shell Petroleum Development Company of Nigeria Ltd (SPDC), was incorporated in Nigeria, and engaged in oil exploration and production in Ogoniland. According to the complaint, after concerned residents of Ogoniland began protesting the environmental effects of SPDC's practices, respondents enlisted the Nigerian Government to violently suppress the burgeoning demonstrations. Throughout the early 1990s, the complaint alleged, Nigerian military and police forces attacked Ogoni villages, beating, raping, killing, and arresting residents and destroying or looting property. Petitioners further allege that respondents aided and abetted these atrocities by, among other things, providing the Nigerian forces with food, transportation,

whether petitioners have stated a proper claim under the ATS, but whether a claim may reach conduct occurring in the territory of a foreign sovereign', responding clearly that '[r]espondents contend that claims under the ATS do not', relying primarily on a canon of statutory interpretation known as the presumption against extraterritorial application. In order to decline the appeal, the Court decided to move a few steps further back from its position in *Sosa*. The decision in Kiebel is determinant:

(1) The principles underlying the presumption against extraterritoriality thus constrain courts exercising their power under the ATS.
(2) To rebut the presumption, the ATS would need to evince a 'clear indication of extraterritoriality'. (*Morrison*, 561 U.S). It does not.
(3) Nothing in the text of the ATS suggests that Congress intended causes of action recognized under it to have extraterritorial reach.
(4) Under the transitory torts doctrine, however, 'the only justification for allowing a party to recover when the cause of action arose in another civilized jurisdiction is a well founded belief that it was a cause of action in that place' (*Cuba R. Co. v. Crosby*, 222 U.S. 473).
(5) In the end, nothing in the text of the ATS evinces the requisite clear indication of extraterritoriality.
(6) Nor does the historical background against which the ATS was enacted overcome the presumption against application in the territory of another sovereign.

---

and compensation, as well as by allowing the Nigerian military to use respondents' property as a staging ground for attacks. Following the alleged atrocities, petitioners moved to the United States where they were granted political asylum and reside as legal residents. Eventually, they filed suit in the United States District Court for the Southern District of New York, alleging jurisdiction under the Alien Tort Statute and requesting relief under customary international law. The District Court dismissed the first, fifth, sixth, and seventh claims, reasoning that the facts alleged to support those claims did not give rise to a violation of the law of nations. The court denied respondents' motion to dismiss with respect to the remaining claims, but certified its order for interlocutory appeal pursuant to §1292(b). The Second Circuit dismissed the entire complaint, reasoning that the law of nations does not recognize corporate liability. The Supreme Court granted certiorari to consider that question. After oral argument, the Supreme Court directed the parties to file supplemental briefs addressing an additional question: 'Whether and under what circumstances the [ATS] allows courts to recognize a cause of action for violations of the law of nations occurring within the territory of a sovereign other than the United States'. After the oral argument, the Supreme Court confirmed the Court of Appeal's decision and gave an answer to the ad hoc question.

Consequently, to the US Supreme Court 'there is no indication that the ATS was passed to make the United States a uniquely hospitable forum for the enforcement of international norms'. As Justice Story put it: 'No nation has ever yet pretended to be the *custos morum* of the whole world'.[120] The Court then added that '[n]othing about this historical context suggests that Congress also intended federal common law under the ATS to provide a cause of action for conduct occurring in the territory of another sovereign. (...) Indeed, far from avoiding diplomatic strife, providing such a cause of action could have generated it. (...) Moreover, accepting petitioners' view would imply that other nations, also applying the law of nations, could hale our citizens into their courts for alleged violations of the law of nations occurring in the United States, or anywhere else in the world. The presumption against extraterritoriality guards against our courts triggering such serious foreign policy consequences, and instead defers such decisions, quite appropriately, to the political branches'. In the end, the Court concluded that the presumption against extraterritoriality applies to claims under the ATS, that nothing in the statute rebuts that presumption ('[t]here is no clear indication of extraterritoriality here'), and that '[c]orporations are often present in many countries, and it would reach too far to say that mere corporate presence suffices. If Congress were to determine otherwise, a statute more specific than the ATS would be required'.[121]

10.2 *Pro-Active Tendencies: Spain and The Prestige*

The Prestige was a Greek-operated, single-hulled oil tanker, officially registered in the Bahamas, but with a Liberian-registered single-purpose corporation as the owner. On 13 November 2002, while the Prestige was carrying a 77,000 metric tons cargo of two different grades of heavy fuel oil, one of its twelve tanks burst during a storm off Galicia in northwestern Spain. Fearing that the ship would sink, the captain called for help from Spanish rescue workers, with the expectation that the vessel would be brought into harbor. Local authorities denied the solicitude and ordered the captain to steer the embattled ship away from the coast and head northwest. After heading towards French water, the vessel was once again forced to change its course and head southwards into Portuguese waters in order to avoid endangering France's southern coast. Fearing for its own shore, the Portuguese authorities promptly ordered its navy to intercept the ailing vessel and prevent it from approaching further. With the French, Spanish and Portuguese governments refusing to allow the ship to

120 *United States v. The La Jeune Eugenie*, 26 F. Cas. 832, CC. Mass., 1822.

121 *Kiobel* was object of special debate in The American Journal of International Law, Vol 106, No 3, 2012.

dock in their ports, the integrity of the single-hulled oil tanker was deteriorating quickly and soon the storm took its toll. It was reported that a section of the starboard hull had broken off, releasing a substantial amount of oil.[122]

On 19 November 2002, the ship split in half, sinking during the same afternoon and releasing over 20 million US gallons (76,000 m^3) of oil into the sea. The oil tanker was reported to be about 250 kilometers from the Spanish coast at that time. An earlier oil slick had already reached the coast. The Greek captain of the Prestige, Apostolos Mangouras, was taken into custody, accused of not co-operating with salvage crews and of harming the environment. After the sinking, the wreck continued leaking oil. It leaked approximately 125 tons of oil a day, polluting the seabed and contaminating the coastline, especially along the territory of Galicia.[123]

The Spanish court understood that the callable liability could not refer to the serious economic consequences of the Prestige oil spill. In addition, it affirmed:

> Articles 109, 110, 116 and related provisions of the Criminal Code, as was apparent from the wording of these provisions when referring to damages caused by the act or the need for any damages arising from the crime, so if accredited only considered the crime of disobedience, not derived from the same or the same have not caused any damage arising from the Prestige oil spill, regardless of the partially good accreditation of huge damages, requiring ad hoc narrow criteria as follows.[124]

Precisely, regarding the criteria to set the terms and scope of liability, the Spanish tribunal stated that 'ex liability offense [had] no other specialty that originated and therefore the Court [could] freely in accordance with proven at trial, set the terms and scope'; nevertheless, 'the rules to be applied for this determination [could] not ignore the fundamental criteria for determining liability in our system , which [meant] that despite the freedom of opinion of

---

122 See Antecedentes de Hecho in *Denuncia de delitos contra el medio ambiente y daños en espacios naturales protegidos. Buque Prestige* – Audiencia Provincial Sección Núm 1. La Coruña (España) – 12/11/2013 (Nº de Recurso: 38/2011).

123 Ibid.

124 See Fundamentos de Derecho (6to Fundamento) in *Denuncia de delitos contra el medio ambiente y daños en espacios naturales protegidos. Buque Prestige* – Audiencia Provincial Sección Núm 1. La Coruña (España) – 12/11/2013 (Nº de Recurso: 38/2011).

what truly [was] accredited in this matter in no case [could] the rules violate disciplined rules of supervision of certain entities and corporations'.[125]

In addition, the Spanish court added that 'laws or contractual agreements limiting certain responsibilities should prevail, so that the entities bound by these rules or contracts [complied] within the limits of such standards and agreements, notwithstanding that may recourse against other obligors'.[126] In that case:

> [T]he current legislation [called] for a series of legal navigation requirements and assurances that the Prestige had at least as essential, but which [were] certainly insufficient to deal with as dire as those consequences caused during the sinking of the Prestige. Precisely, why they have created organisms that may get compensation tighter as with the IOPC Fund, which has some limitations of compensation agreed by and with international criteria linking many States including the Spanish, then these constraints must be respected, not so much to prevent a dramatic disappearance of these organisms as to comply with limitations in the art that are the essence, or in other words, if there are regulations limiting such compensation against certain organisms, they must be respected and demand of financial liabilities not satisfied by these other agencies responsible to this limitation may not apply.[127]

The reason given by the Spanish tribunal to deny liability encompassed the relation and proportionality between the activities, the risks, and the liability.[128]

Regarding the responsibility of other agencies and private entities involved in the maritime activity, the Spanish tribunal declined to bring to court either of the mother companies or to turn down the existent liability fund limit.[129]

---

125 Ibid.

126 Ibid.

127 Ibid.

128 'It could be argued that the uniqueness of the business that took place in and through the Prestige allows your confusing management then listed as responsible legal persons of low economic power and quite unable to cope with such responsibility, which may be true and if it is can tolerate in their own field of business, it would be even possible to go to the doctrine of piercing the corporate veil to identify the real culprits'. See Fundamentos de Derecho in *Denuncia de delitos contra el medio ambiente y daños en espacios naturales protegidos. Buque Prestige* – Audiencia Provincial Sección Núm 1. La Coruña (España) – 12/11/2013 (Nº de Recurso: 38/2011).

129 Ibid.

11 Conclusions

Technology has always evolved faster than legal frameworks whilst industrial activities have expanded offshore at an increasingly overwhelming rate. International environmental damage regulation should start to foresee the necessity of a change in the jurisdictional approach before it is far too late.[130] Meanwhile, scholars seem to oscillate in the debate between 'apology' and 'utopia' regarding the materialization of the principle of CHM.[131]

Professor Baslar suggests that 'the view that the difficulty of bringing the common heritage of mankind into the corpus of international law may be surmounted by having recourse to the Grotian or eclectic philosophy'.[132] In the end, 'the common heritage of mankind as a distinctive illustration of idealism relies on different sources like *inter alia*, resolutions and declarations of international organizations, quasi-legislative acts of the United Nations and *jus cogens*'.[133] In this context, an Opinion rather *démodé* was delivered by the SDC withholding strong State-centered criteria and upholding similar reticence against 'universal jurisdiction' (*actio popularis* and *locus standi*) as some classical and contemporary national jurisprudence. Even though an opportunity to make the sharper statement that both UNCLOS and the ITLOS are independent guardians of the security of oceans' environment might have passed by, and bearing in mind that the general panorama is not a positive one,[134] tendencies may change – sooner rather than later.[135]

---

130 See J.N. Moore and M.H. Nordquist, *Current Marine Environmental Issues and the International Tribunal for the Law of the Sea*. Martinus Nijhoff, 2001.

131 'Positivists, on the one side, apologize that the traditional sources of international law do not give due credibility to the common heritage of mankind as a legal principle. Whereas, on the other side, non-positivits claim that the common heritage of mankind is an established legal principle of international law, acquiring its legal normativity from other sources such as general principles of law and even from peremptory norms'. Fn. 6, at p. 367.

132 Ibid., p. 368.

133 Ibid., p. 359.

134 See X. Yang, *State Immunity in International Law*. Cambridge University Press, 2012.

135 See D. Vidas (ed), *Law, Technology and Science for Oceans in Globalisation*. Martinus Nijhoff, 2010; also, S. Kaye, 'The Law of the Sea, Marine Technology and Global Social Justice', in A. Chircop, T.L. McDorman and S.J. Rolston (eds), *The Future of Ocean Regime-Building*. Martinus Nijhoff, 2009.

Bibliography

Bergkamp, L., *Liability and Environment*, Kluwer Law International, 2001.

Birnie, P.W. and Boyle, A.E., *International Law and the Environment*, Clarendon Press, 1992.

Faure, M.G., Lixin, H. and Hongjun, S., *Maritime Pollution Liability and Policy*, Wolters Kluwer, 2010.

Flauss, J.-F., 'Competence civile universelle et droit international general', in C. Tomuschat and J.M. Thouvenin (eds), The Fundamental Rules of the International Legal Order, Nijhoff, 2006.

Gould, H., *The Legacy of Punishment in International Law*, Palgrave Macmillan, 2010.

Karavias, M., *Corporate Obligations Under International Law*, Oxford University Press, 2014.

Kaye, S., 'The Law of the Sea, Marine Technology and Global Social Justice', in A. Chircop, T.L. McDorman and S.J. Rolston (eds), *The Future of Ocean Regime-Building*, Martinus Nijhoff, 2009.

Larsson, M.L., *The Law of Environmental Damage*, Kluwer Law International, 1999.

Lillich, R.B., '*Sovereignty and Humanity: Can They Converge*?' in A. Grahl-Madsen, J. Toman and W. de Gruyter (eds), The Spirit of Uppsala: Proceedings of the Joint UNITAR-Uppsala University Seminar on International Law and Organization for a New World Order, Uppsala, 1984.

Maslar, K., *The Concept of the Common Heritage of Mankind in International Law*, Martinus Nijhoff, 1998.

Moore, J.N. and Nordquist, M.H., *Current Marine Environmental Issues and the International Tribunal for the Law of the Sea*, Martinus Nijhoff, 2001.

Muchlinski, P.T., *Multinational Enterprises and the Law* (Oxford International Law Library, 2007.

Nijman, J.E., *The Concept of International Legal Personality*, Asser Press, 2004.

Nolte, G., 'Universal Jurisdiction in the Area of Private Law – The Alien Claims Act', in C. Tomuschat and J.M. Thouvenin (eds), The Fundamental Rules of the International Legal Order, Nijhoff, 2006.

Portmann, R., *Legal Personality in International Law*, Oxford University Press, 2010.

Post, A.M., *Deepsea Mining and the Law of the Sea*, Martinus Nijhoff, 1983.

Redgwell, C., 'From Permission to Prohibition: The 1982 Convention on the Law of the Sea and Protection of the Marine Environment' in D. Freestone, R. Barnes, and D. Ong (eds), The Law of the Sea: Progress and Prospects, Oxford University Press, 2006.

Vidas, D. (ed), *Law, Technology and Science for Oceans in Globalisation*, Martinus Nijhoff, 2010.

Yang, X., *State Immunity in International Law*, Cambridge University Press, 2012.

CHAPTER 11

OSPAR and Coastal State Encroachment on High Seas Submarine Cable Freedoms

Douglas R. Burnett

1 Introduction

For over 166 years, submarine cables have brought the world closer together. But the advent of fibre optic technology has turned submarine cables into a critical international infrastructure upon which the world's digital economy and the internet depend.

The first part of this paper examines how cables are integral to sustainable development. This development has taken place with a neutral impact on the marine environment. The second part of the paper examines recent actions taken by the Convention for the Protection of the Marine Environment of the North-East Atlantic (OSPAR) and argues that extending these or similar actions to regulate submarine cables on the high seas undermines the United Nations Law of the Sea Convention (UNCLOS)[1] and imposes needless risks of unintended consequences on the world's critical international submarine cable infrastructure. The third part of the paper examines legal and environmental aspects of cables and the adequacy of existing ocean governance on the high seas proper and in the Area. Central to the success of submarine cables has been the freedom to the laying and repair of cables under UNCLOS and customary international law. Coastal State encroachments, current discussions on biodiversity beyond national waters and a possible new treaty to regulate international cables on the high seas proper and the Area as defined in UNCLOS threaten to undermine the current successful balance between submarine cables and protection of the marine environment.

In this paper, 'BBNJ' refers to the high seas proper and the Area beyond national jurisdiction as defined in UNCLOS. Additionally, 'submarine cable community' refers to cable owners, cable ship operators, marine route surveyors, scientific institutions and interested governments.

1 United Nations Convention on the Law of the Sea (adopted 10 December 1982, entered into force 16 November 1994) 21 ILM 1261 (1982) (UNCLOS).

 | DOI 10.1163/9789004360273_012

2 Submarine Cables are Critical International Infrastructure Crucial for Development

The Oceans and the Law of the Sea Report of the Secretary-General[2] succinctly sums up the conventional wisdom about international submarine cables and sustainable development:

> 53. Submarine cables are critical communications infrastructure, being used for more than 98 per cent of international internet, data, and telephone traffic, with only a few States being without fibre connectivity, and many of these having cable projects currently underway.[3] Submarine cables are recognized as vitally important to the global economy[4] and hence to economic growth. By underpinning international communications, their role in providing access to data and information for all people is evident. [...]
>
> 55. Functioning as the backbone of the international telecommunications system, submarine cables are directly part of the global critical infrastructure and sustainable industrialization and indirectly they contribute to all other areas recognized as important for sustainable development.

Amazingly, when people think about international communications, they often mistakenly regard satellites as the primary medium of modern international communications. They express surprise to learn that over 98% of international communications are carried by a relatively small number of fibre-optic submarine cables with diameters akin to a domestic garden hose even though this has been the case for almost 30 years. The confusion is understandable. The idea that a person's cell phone air link is sent to a nearby cell towerand that the overseas messages themselves are then broken down into bits of data, which then at the speed of light ply the ocean depths on unseen cables, is hard to imagine. This is difficult to comprehend. The tremendous volume of data carried at low cost by modern fibre-optic submarine cables dwarfs the limited capacity of higher cost satellites. For example, the capacity

---

2 UNSG Report A/70/74 (30 March 2015).

3 D. Burnett, D. Freestone, and T. Davenport, 'Submarine Cables in the Sargasso Sea: Legal and Environmental Issues in Areas Beyond National Jurisdiction', Report from the Workshop held 23 October 2014 at George Washington Law School, Washington, D.C. (16 January 2015), <https://www.iscpc.org/documents/?id=1792> accessed 1 July 2017.

4 UNGA Res. 69/245 (29 December 2014).

of a single transatlantic cable has increased by a factor of 100,000 in 25 years.[5] Additionally, the technical transmission delays, modest capacity and other quality limitations inherent in satellites make them comparatively marginal for continuous transmission of high speed voice, video, and data traffic.

The collective impact of the laypersons' mistaken beliefs and knowledge gap is negatively compounded by the fact that many in government share their misconceptions, even as they fashion ocean policies and regulations that overlook submarine cable history, marine engineering, seamanship, environmental aspects and international law. Not infrequently, these flawed regulatory efforts undercut the viability of the successful submarine cable network as the critical international infrastructure upon which the internet and global economy are based. Even more popularly unknown or appreciated is the substantial body of scientific research and records that document the inter-relationship of submarine cables and the marine environment. Like cables, the time tested and very successful international legal regime that supports international cables is often misunderstood or overlooked.

2.1 *Fibre Optic Submarine Cables*

Each day the Society for Worldwide Interbank Financial Telecommunications (SWIFT) transmits 15 million messages over cables to over 8,300 banking organizations, security institutions and corporate customers in 208 countries. The Continuous Linked Settlement (CLS) Bank located in the United Kingdom is just one of the critical market infrastructures that rely on SWIFT as it provides global settlement of 17 currencies with an average daily US dollar equivalent of approximately USD 3.9 trillion. The U.S. Clearing House Interbank Payment System (CHIPS) is another structure that processes over USD 1 trillion a day to over 22 countries for investment companies, security and commodity exchange organizations, banks and other financial institutions.[6]

If the approximately 40 or so garden hose diameter cables connecting the United States to the rest of the world were cut, even using every single satellite in the sky, it is estimated that only 7% of the total United States traffic volume could be carried by satellite.[7] Referring to the submarine cable networks, the Staff Director for Management of the Federal Reserve observed that 'when the

5 J. Chesnoy, 'Back Reflection', *Subtel Forum Magazine*, Issue 88, 2016, p. 96.

6 S. Malphrus, 'Undersea Cables and International Telecommunications Resiliency', 34th Annual Law of the Sea Conference, University of Virginia, May 2010.

7 D. Burnett, Testimony before the Senate Foreign Relations Committee on the United Nations Law of the Sea Convention, October 2007.

communication networks go down, the financial sector does not grind to a halt, it snaps to a halt.'[8] The same can be said for most industries enmeshed in the global economy through the internet including shipping companies, airlines, banks, supply chains, manufacturing businesses, and entertainment industries.

Other countries are no different in their reliance. Australia and Singapore, for example, each rely on several cables landing in each nation for over 99% of their international communications. Japan does the same with about 20 international cable systems. And the list goes on. With the laying of submarine cables along the east coast of Africa in 2009–2010, the last major group of nations now has access to the world's submarine cable network. As of mid-2012, only 22 nations and territories remained isolated from fibre-optic connectivity and many of these have connecting cable projects underway.[9] A major challenge now being met by the submarine cable community is providing connections to small island economies and isolated coastal communities of the high Arctic, together with the provision of redundant cable connections to keep these economies connected in the event of a cable fault.

The world's dependence on reliable low cost and secure submarine cables continues to grow. 'Every *second* they can carry 31 terabits across the Pacific and 55 terabits across the Atlantic.'[10] A look at the websites of major companies like Google, Microsoft, Facebook and Amazon shows the diverse locations of the legions of computer servers in each company's data centres which are distributed worldwide and on every continent except Africa. These *cloud* data centres are seamlessly connected by international submarine fibre-optic

---

8 S. Malphrus, Board of Governors of the Federal Reserve System, First Worldwide Cybersecurity Summit, Dallas, Texas, 3–5 May 2010.

9 Submarine Telecoms Forum Inc., 'Telecoms Industry Report', 2012, pp. 14–15. Inhabited sovereign States and territories without fibre optic connectivity include: Somalia, Saint Helena, Ascension, and Tristan da Cunha (British Overseas Territory), Christmas Island (Australian External Territory), Montserrat (British Overseas Territory), Saint Pierre and Miqulon (French Collectivité d'outre-mer), Easter Island (Chilean Special Territory), Falkland (Malvinas) Islands (British Overseas Territory), Cook Islands (Self-Governing State in Free Association with New Zealand), Kiribati, Nauru, Niue (Self-Governing State in Free Association with New Zealand, Norfolk Island (Australian External Territory), Palau, Pitcairn Islands (British Overseas Territory), Solomon Islands, Tokelau (New Zealand Dependent Territory), Tonga, Vanuatu, Wallis and Futuna (French Collectivité d'outre-mer), Timor Leste.

10 The Economist, 'The See-Through Sea', 16 July 2016, p. 16.

cables. It is not an exaggeration to say the *cloud* would not exist but for cables under the sea.

By 2020, one expert estimates that there will be 4 billion people connected to the internet, USD 4 trillion of revenue opportunity from these connections, using over 25 million apps, with over 25 billion embedded intelligent systems, and 50 trillion gigabits of data.[11] These connections will exist almost exclusively on international submarine cables, the backbone of the internet. Applications known to many such as Skype, Facetime, Netflix, Twitter, Facebook, and YouTube remind us all in a personal way that our lives are directly impacted by submarine cables. The unparalleled ability of submarine cables to increase GDP is also recognized by the World Bank:

> Subsea Fiber has recently gained renewed focus within the TMT Investment sector of IFC (the private investment arm of the World Bank); as it has a broad economic impact on developing economies. Subsea Fiber provides increased international capacity, which usually equates to a drop in wholesale pricing, and open access to service providers. As volume and completion increases, prices fall to the enterprise and the end consumer. This will stimulate development of new business models regarding health, education, and commerce. Additionally, increased subscriber rates due to access to low cost, high speed internet access is one of the key catalysts to economic development. The World Bank estimates that a 10% increase in Broadband Internet Access contributes to an increase of 1.38% in GDP.[12]

Affordable telecommunication services are now a reality in places where internet connectivity was nearly unheard of just a few years before. The new systems in Africa and Asia-Pacific are prime examples showing the economic impact that just one fibre-optic cable can have. World Bank press releases herald these transformative cable impacts:

> This cable is more than simply an important piece of technology. It is a key infrastructure project that can deliver extraordinary benefits across Samoa's education, health, business and tourism sectors for decades to come.[13]

11 Source: Mario Morales, IDC.

12 L. Cook, cited in *Submarine Telecoms Industry Report*, vol. 4, 2015, p. 15.

13 F. Dees-Gross, 'Underwater Cable to Bring High Speed Internet to Samoa', quoted in The World Bank Press Release, 19 June 2015.

> Faster, cheaper, and more reliable connections can result in the development of:
>
> - New opportunities to share information: teachers, doctors, farmers, and fishermen use technology to communicate, share information, buy goods, find better prices, make payments, improve the reach of their services and increase their bargaining power.
> - E-services: Developed by Pacific governments to provide geographical information systems, new modes of distance learning, and online business applications.
> - Trans-national cooperation: On issues such as monitoring natural resources like fisheries, disaster mitigation, and collaboration on service deliver like health and education, including in remote areas.
>
> There is huge potential to harness the power of technology to create economic growth and opportunities to reduce poverty.[14]

Africa is also a bellwether example of the transformative impact of modern fibre-optic submarine cables:

> A key indicator for broadband development in Africa is the deployment of basic infrastructure, such as international submarine cables. Many African governments have co-sponsored new cables of this type along the continent's east and west coasts, with the aim of improving broadband connectivity.[15]

Other examples illustrate the wide use and importance of cables to modern life. The National Marine Fisheries Service of the United States tracks trade in fish products between the US and China valued at USD 100 million per annum and these transactions are carried out on submarine cables.[16] Through cables, unique environmental and cultural tourism in small developing island communities like Vanuatu and Fiji are promoted. The impact is demonstrated by a single group in Fiji that employs over 600 persons in a diverse portfolio of tourist related services including transportation, retail travel, overseas wholesaling, and hotel ownership and development. International telemedicine at

14 The World Bank, 'An Information-communication Revolution in the Pacific', 18 May 2013.

15 The Broadband Commission for Digital Development, 'Broad Band: A Platform for Progress', 2011, p. 155.

16 <www.st.nmfs.noaa.gov/commercial-fisheries/foreign-trade> accessed 1 July 2017.

world-renowned facilities in Dubai and Minnesota is a reality through cables worldwide. Cables allow instantaneous machine-to-machine and artificial intelligence solutions to improve healthcare on many levels.

The future contributions of submarine cables to sustainability are bright and varied. Technology advances in 3D printing will allow items to be locally manufactured, reducing the need for ocean or air transportation and other high carbon footprint industries. For example, an electric lightweight 80 km/h motorcycle entered print production this year.[17] All that will be required is an international fibre-optic cable connection to allow the software, blue prints, and payments to flow. Similarly, long distance learning will allow disbursed populations in archipelagic States to have common access to the best teachers and resources. Medical data and expertise will allow for improved health care and bring high-level care to more and more people. Mapping and location services enabled by cables will save energy, time, and funds in providing government and private industry services with the precision needed for their efficient management and provision.

2.2 *Submarine Cables and Science*

Besides the critical economic and social sustainability roles played by the submarine cables as international arteries connecting the world, submarine cables are also vital for marine science and the quest to learn more about the oceans and climate. In a 2009 survey, the ICPC identified 193 ocean observation sites and areas worldwide, including at least 34 that planned or were using submarine cables for data transmission and power transfer in the world's oceans.[18] The 800 km cable-based Neptune system, with multiple scientific nodes (special seabed housings capable of supporting monitoring equipment and experiments) off Vancouver Island, British Columbia is a standout operational example. Another is the Ocean Observatories Initiative (OOI) array off Oregon, which like its Canadian counterpart is based on a 900 km cable that supports a suite of nodes. Japan has pioneered the use of submarine cable systems to monitor seismic activity and detect tsunamis.[19]

---

17 The Economist, 'Alloy Angels', 28 May 2016, p. 72.

18 ICPC Ocean Observation Sites and Areas, 2009, see <www.iscpc.org> accessed 1 July 2017. The survey results were compiled by Lionel Carter, Victoria University, Wellington, New Zealand, the ICPC International Marine Environmental Advisor (IMEA).

19 C. Manoj et al., 'Can Undersea Voltage Measurements Detect Tsunamis?', *Earth Planets Space*, vol. 58, 2006, pp. 1–11; R. Monastersky, 'The Next Wave', *Nature*, vol. 483, 2012, pp. 144–146.

2.3 *Submarine Power Cables and* ABNJ

Although legally holding the same status, there are important differences between power cables and telecommunication cables.[20] However, due to physical depth, weight and length limitations, no power cables have been laid or are planned to be laid in the high seas. A power cable discussed between Iceland and the United Kingdom, if realized, will be laid along the continental shelf and the Exclusive Economic Zone (EEZ) of those States and the Faroe Islands – and not in the Area Beyond National Jurisdiction (ABNJ) – in order to keep the required cable depth for laying the cable to less than 1000 metres.[21] For reference, a comparison chart of differences between submarine telecommunication and High Voltage Direct Current (HVDC) submarine power cables in the ABNJ is attached as Annex A.

The physical size of submarine power cables in the ABNJ is unknown as there are currently no examples that extend into that area. However, on the continental shelf and the EEZ, diameters range from typically 80 mm to 150 mm. Telecommunication cables in the BBNJ are typically 17–22 mm because water depths minimise threats from bottom trawling and shipping, and hence the need for protective armour. Power cables are associated with electromagnetic fields (EMF) that are constrained closely to the cable.[22] Our present knowledge of any effect of power cable EMF on marine organisms, while incomplete, concludes that power cable fields have no negative effect on the marine organisms studied, which includes animals sensitive to such fields.[23]

---

20 M. Eccles, J. Ferencz, and D. Burnett, 'Submarine Power Cables', in D. Burnett, R. Beckman and T. Davenport (eds.), *Submarine Cables: The Handbook of Law and Policy*, Martinus Nijhoff Publishers, 2014, pp. 301–322.

21 H. Hafsteinsdót, 'Presentation on the Iceland-UK HVDC Cable', at the workshop 'Legal Status of Submarine Cables, Pipelines and ABNJ', Ankara, Turkey, 7–8 April 2016, scheduled for publication, available at dehukam@ankara.edu.tr.

22 Normandeau Associates, Inc. et al., *Effects of EMFs from Undersea Power Cables on Elasmobranchs and Other Marine Species*, published by US Dept. of the Interior, 2011.

23 A. Copping et al., 'State of the Science Report: Environmental Effects of Marine Renewable Energy Development Around the World', Annex IV, 2016, <https://www.ocean-energy-systems.org/.../39055-state-of-the-science-report-2016.pdf/> accessed 1 July 2017.

3 OSPAR's Approach to Regulate Submarine Cables on the High Seas

3.1 *OSPAR High Seas Marine Protected Areas*

OSPAR has declared nine marine protected areas (MPAs) on the high seas. Concerns have been raised by the submarine cable community when OSPAR unilaterally and without any consultation with ocean submarine cable stakeholders declared these high seas MPAs over five existing transatlantic cable systems.[24] However to date the declaration appears inactive for submarine cables with no impact or restrictions on the laying and repair of these systems.[25]

If a high seas MPA were to become a 'no go' or restricted area for international submarine cables or if repairs to existing cables were delayed or prejudiced, then of course the cable owners and/or cable ship operators would request their respective States to take up the issue diplomatically to prevent endangering this global critical infrastructure.

3.2 *OSPAR's Best Environmental Practice on Laying and Operation of Cables*

The core of the submarine cable community concern is that, coincidently with the declaration of high seas MPAs, OSPAR has also unilaterally and without consultation of stakeholders issued a document entitled 'Guidelines on Best Environmental Practice (BEP) in Cable Laying and Operation' (Agreement 2012/2). Because BEP was crafted by OSPAR policy makers with no consultation – or even awareness – of national members of the submarine cable community such as national telecom companies, cable ship operators etc., the BEP is seriously flawed with fundamental errors that have caused deep mistrust and wariness within the submarine cable community. Some of the more striking points in BEP are in italics and summarized below. This listing is not exhaustive, but it is sufficient to show the need for changes to the BEP to consider facts overlooked, misunderstood, or wrong. The positive argument is made that a workshop(s) involving OSPAR representatives and the submarine cable community, including scientists, engineers, mariners and governments may be the best means for allowing for productive dialogue, information sharing and thoughtful review of the BEP. A fair and frank exchange of information

24 Ibid., pp. 208–212.

25 The ICPC now has observer status with OSPAR and is working to challenge and educate OSPAR about submarine cables and accurate considerations for best environmental practices.

and views by the submarine cable community and OSPAR policy makers can dramatically improve the current negative environment for cooperation. A continuation of what is perceived to be an arrogant 'top down' approach by OSPAR will heighten the current mistrust and stir opposition.

- *The BEP erroneously focuses on power cables filled with oil insulation and suggests that such cables are commonly used now for submarine cables in the oceans.* International submarine power cables discontinued the use of oil for insulation in cables in the 1990's, because of environmental and maintenance concerns and adopted for insulation mass impregnated paper or the more common cross linked polyethylene (XPLE).[26] There are no oil-filled submarine cables in use or contemplated for international submarine power cables. This fact would be especially true in ABNJ, but, as noted in the preceding section and Annex A, there are no HVDC power cables in ABNJ and none are forecast. So the BEP's regulation of HVDC cables in ABNJ is a solution looking for a problem.
- *Contrary to BEP assumptions, submarine cable repairs in ABNJ are rare.* Statistics for international telecommunication cable repairs for 2008–2015 show that in the ABNJ, there are only on average about four cable repairs per year *worldwide*. While new systems will be laid from time to time, the expense and technical expertise required to lay and operate these systems limits the number of systems successfully installed. Cable capacity upgrades that increase the capacity of existing cables by simply changing equipment at cable landing stations, without the need for laying a new cable, have significantly reduced cable laying worldwide, extending considerably the approximately 25-year design commercial life of existing cables.
- *Route selection for cables in ABNJ requires a formal approval process (or several if necessary) with integrated environmental impact assessment (EIA).* As discussed in the third part of this paper, this requirement is not accurate and is inconsistent with express provisions of UNCLOS that do not allow coastal States outside of territorial seas, in the EEZ, on the legal continental shelf or the high seas to delineate cable routes or impose conditions except for cables used for the exploitation and exploration of natural resources. With regard to EIAS, as discussed in this paper, there is existing authority for coastal States or flag States to require an EIA if there is a substantial threat

26 See D. Burnett, R. Beckman and T. Davenport (eds.), *Submarine Cables: The Handbook of Law and Policy*, Martinus Nijhoff Publishers, 2014, p. 312.

of environmental harm.[27] The fact that EIAs on fibre optic cable systems are rare outside of territorial seas speaks to the scientific evidence in peer review studies, literature, and workshops that demonstrate such systems are environmentally benign or neutral. Some of this large amount of literature is highlighted for ease of reference in Annex B and the footnotes.

- *Protected areas should be established in ABNJ where cable laying would be restricted.* The OSPAR designation of nine high seas MPA's without notice to the submarine cable community, enveloping six active fibre optic cable systems owned or partially owned by foreign companies not incorporated in OSPAR State parties,[28] and BEP attempts to regulate the laying and repair of submarine cables in these areas is inconsistent with express provisions of UNCLOS described in this paper. The BEP also has no provision to grandfather existing cables, which is not consistent with international law and practice for submarine cables and imposes an unreasonable economic cost not capable of being foreseen by the owners of these cables when laid.[29] Furthermore, no evidence or study by OSPAR has been presented to show that cables would have a negative impact on any MPA and hence should be subject to any ABNJ restrictions. To the contrary, as previous noted, cables have a neutral effect on the marine environment (Annex B, e.g. references 11–13).
- *Submarine cables should be laid in ABNJ by 'bundling with existing cables and pipelines'.* Besides being inconsistent with express provisions of UNCLOS, this requirement increases the risk of marine oil pollution of the marine environment from pipelines by forcing cable vessels to grapnel during repairs near pipelines. The proposal also increases the likelihood of cable faults by ignoring reliable and safe practices that have stood the proven test of time over the past 166 years. Cables are not laid closely next to other cables. The historically proven practice is to lay cables with at least a separation distance of three to five times the water depth to allow for safe recovery in case of repair. Cable ship operators and the cable owners who charter these ships would not engage in imprudent actions that could put at risk specialist operations coverage under customarily carried marine insurance.

27 Article 206 of UNCLOS.

28 International fiber optic cable systems in OSPAR high seas MPAs: Atlantic Crossing, TAT-12/13, TAT-14, Hibernia, Apollo, and FLAG.

29 Article 51 of UNCLOS is a useful example of the use of grandfathering for existing submarine cables. In this case international cables laid in what had previously been considered to be high seas and became archipelagic waters. This provision allowed the existing cables to continue to be replaced and maintained without commercial impact.

- *Minimize by regulation cable and pipeline crossings in ABNJ*. Besides being inconsistent with the express provisions of UNCLOS, this requirement is unnecessary. There are well tested and understood crossing practices[30] routinely used that allow cables to cross previously laid cables and pipelines. These techniques have been employed on hundreds of occasions with few problems and no record of marine pollution.[31] This is a solution looking for a problem that does not exist.
- *Regulatory cable route re-routing based on EIS in the ABNJ*. As described in this paper, this requirement is inconsistent with UNCLOS express provisions that do not allow delineation of cable routes by coastal States in the EEZ, on the continental shelf and most certainly on the high seas – nor is there a need demonstrated for such impairment to the freedom to lay submarine cables on the high seas.
- *Regulation of anchor arrays used by cable ships in the ABNJ, including using 'tenders to lift anchors rather than dragging them across the seabed.'*[32] This is not a practice used in laying and repair of international submarine cables in the ABNJ. In the ABNJ, the working depths in thousands of meters render anchor use impracticable in any event. In other waters, modern lay vessels do not operate anchors but use different types of Dynamic Positioning (DP) systems with no anchor use at all. Even in shallow territorial waters where a barge may be used, anchors are not dragged over the sea floor, but are precisely positioned by anchor handling tugs.
- *Regulation of cable burial in ABNJ*. In ABNJ, submarine cables are not buried, but the garden hose-diameter cable is simply laid on the seabed, following its contour. Cable burial, even if technically and economically possible at the depth where cables are laid in the ABNJ (which it is presently not normally over 1000 meters, or in a few exceptional cases, up to 1500 meters water depth), serves no purpose and would potentially harm the marine biota for no reason since the cables at those depths require no burial protection.
- *Regulation by encouraging horizontal drilling in ABNJ*. This is not a high seas practice nor is it realistically feasible outside of territorial seas and certainly not in ABNJ.

---

30 See ICPC, 'Telecommunications Cable and Oil Pipeline/Power Cable Crossing Criteria', Recommendation No. 10A, 2011; D. Burnett, R. Beckman and T. Davenport (eds.), *Submarine Cables: The Handbook of Law and Policy*, Martinus Nijhoff Publishers, 2014, pp. 214–215 and Figure 8.1.

31 For example a recent laying of a fiber optic cable in the Atlantic along the west coast of Africa crossed charted 48 submarine cables and 10 pipelines without incident.

32 BEP 5.2.1.

- *Regulation of the use of rock-mattress covers in the case of crossing of cables and pipelines using inert natural stone.*[33] Mattress use is not a practice used in the ABNJ, but if used, concrete mattresses – not stones – would be used for reasons of practicability. BEP's reference to 'stones' is exaggerated and incorrect, because concrete mattresses are used but only in discrete near shore risk areas where a cable needs protection and cannot be buried because of hard or rocky seabed. In such areas, soft sediments are not present. Since cables are not buried, but simply surface laid in ABNJ, this regulation is not realistic.
- *Underwater noise.*[34] The duration of a cable lay is short and occurs only in the initial hours or day of the lay of a cable with a 25–year service life. The method of cable laying is optimized to reach the target burial depths required to protect the cable from hazards at those depths. The entire process is brief and non-repetitive. Submarine cable operation is a one-off event. 'For most organisms, it is unlikely they will affect more than once during their lifetime. This contrasts with chronic anthropogenic sounds, such as exploration surveys in hydrocarbon provinces, shipping channels, maintenance dredging or fishing activities.'[35] The cost and benefit ratio to change this proven method to minimize a brief transitory noise impact jeopardizes this aim. Submarine cable route survey (side scan survey – high frequency but short range) and engineering operations cover great distances in ABNJ at moderate ship speed. 'For this reason, techniques that are used in civil engineering, such as bubble curtains and coffer dames, are not likely to be cost-effective for submarine cable operations.'[36] Moreover, since cables are not buried in the ABNJ, noise concerns from one off burial operations are not relevant to the marine environment in this deep sea environment.
- *Regulation of the times and routes for cable laying in ABNJ by limiting times based on bird migration, spawning grounds, calving.* As discussed in the next section, this practice is inconsistent with UNCLOS provisions in the ABNJ. Comprehensive studies and experience have shown cable activities

33 BEP 3.2.

34 BEP 3.3.

35 T. Worzyk, *Submarine Power Cables: Design, Installation, Repair and Environmental Aspects*, Springer, 2009.

36 R. Hale, 'Regulation of Underwater Sounds', *Suboptic*, 2016, p. 5.

are not significant issues for whales and sharks.[37] The cable community is unaware of any bird impacts associated with cable laying in ABNJ.

- *Blasting of rocky subsoil and use of explosives in cable laying should be regulated.* Explosives, air guns, etc. are simply not deployed in laying and repair of submarine cables. BEP has no reference as to the origin of such a claim. In fact, such activity has been long recognized as potentially harmful to cables.[38]
- *The '2 K criterion' test to control heat emission should be used by regulators to specify burial depths of 1 to 3.5 meters.* This requirement test is wrong, as demonstrated by well-regarded studies by German scientists in peer reviewed literature.[39] The test is based on an unproven theory created by a biologist that did not use year-round measurements of seabed temperature. As a result, the normal seabed temperature variation by season was not considered. Had it been considered, as the recent studies demonstrate, the '2 K criterion' is not scientifically valid.[40]
- *Regulate removal of out-of-service cables, including EIAs and requirements for fluid filled cables in the ABNJ.* This requirement is inconsistent with UNCLOS and ignores the custom and practice[41] of the cable community with respect to management of out-of-service submarine cables that is consistent with UNCLOS. As indicated, there are no fluid filled cables in use in the ABNJ (nor have there ever been). Furthermore, the BEP prevents out-of-service cables from being stored for re-use for telecommunications and science and

37 L. Carter et al., 'Submarine Cables and the Oceans: Connecting the World', in *UNEP-WCMC Biodiversity*, Series No. 31, 2009, pp. 31–33; D. Burnett, R. Beckman and T. Davenport (eds.), *Submarine Cables: The Handbook of Law and Policy*, Martinus Nijhoff Publishers, 2014, Chapter 7, pp. 184–185, and Annex B.

38 ICPC, 'Procedure To Be Followed Whilst Offshore Seismic Survey Work Is Undertaken In The Vicinity Of Active Submarine Cable Systems', Recommendation No. 8, 2011.

39 C. Müller et al., '2 *K-Criterion, Measuring and Modelling Temperatures and Thermal Connectivities and Diffusivities in Shallow Marine Sediments*', Proceedings of the Conference on Marine Energy at the TUHH Hamburg 2013; C. Müller et al., 'Temperatures in Shallow Marine Sediments: Influence of Thermal Properties, Seasonal Forcing, and Man-made Heat Sources', *Journal of Applied Thermal Engineering*, vol. 108, 2016, pp. 20–29; accessible also at <http://dx.doi.org/10.1016/j.applthermaleng.2016.07.105> accessed 1 July 2017.

40 Ibid.

41 See ICPC, 'Management of Out-of-Service Submarine Cables', Recommendation No. 1, 2011.

recycled for use as artificial reefs.[42] While BEP seems to concede that removal of cables from the marine environment after 25 or more years is likely to cause more harm to the environment than leaving them in place, BEP fails to provide any cost benefit analysis for its crude and flawed mitigation.

- *Environmental data in connection with a cable project should be made publically available, except for confidential commercial information.* This requirement, especially outside of the territorial seas, would subject all submarine cables to the UNCLOS requirements of Marine Scientific Research (MSR) by States others than those where the system lands.[43] This would make the laying of international submarine cables outside of territorial seas impracticable and subject to effectively a veto by non-landing coastal States. It should be noted that the cable community has a long history of supplying data and samples, together with survey information for scientific research to coastal States where the cable lands as part of the permitting process. Out-of-service cables have been re-used to effectively monitor ocean currents and tsunami events. The relationship with organizations involved in marine research continues today at an increasing scale (Annex B). Regulation of these scientific cable activities would stifle innovation and cooperation.
- *Financial payments for mitigation.* This requirement is inconsistent with express freedoms to lay and maintain international submarine cables, especially when imposed by a group of OSPAR States over other non-OSPAR nations and their nationals on the high seas.[44] Needlessly increasing the cost under a toll-like system to lay and repair cables impacts the ability of cables to reach small island developing States and provide critical redundancy to States served by only a single cable.

3.3 *Inconsistencies between OSPAR Best Practices and UNCLOS*

UNCLOS provides with express language for the freedom to lay and maintain international submarine cables in the EEZ,[45] the continental shelf,[46] and the

42 See D. Burnett, R. Beckman and T. Davenport (eds.), *Submarine Cables: The Handbook of Law and Policy*, Martinus Nijhoff Publishers, 2014, Chapter 8 (Out-of-Service Submarine Cables), pp. 214–215.

43 See, generally, T. Davenport, *Science Technology and New Challenges to Ocean Law*, Brill, 2015, Chapter 8 (Submarine Cables and Science: A New Frontier in Ocean Governance), pp. 224–252.

44 See D. Burnett, R. Beckman and T. Davenport (eds.), *Submarine Cables: The Handbook of Law and Policy*, Martinus Nijhoff Publishers, 2014, pp. 150–151.

45 Article 58 of UNCLOS.

46 Articles 78 and 79 of UNCLOS.

high seas.[47] A detailed description of the UNCLOS legal regime can be found in *Submarine Cables: The Handbook of Law and Policy*. Unlike a pipeline, in the EEZ and on the legal continental shelf a coastal State may neither delineate the cable route, nor impede, nor prejudice the possibility of repair of the cable.[48] In the ABNJ, where the coastal State may not restrict the freedom of other States or their nationals to lay and maintain cables, the freedom to lay and maintain cables is limited only to taking measures to avoid the possibility or prejudicing cable repair and due regard for other freedoms of the sea and activities conducted in the Area. The BEP's recommendations discussed above are inconsistent with UNCLOS to the extent they restrict, impede, or fail to show due regard for the express freedom to lay submarine cables and prejudice cable repair.

The concept of a few States unilaterally imposing an MPA on the high seas and then using this authority to restrict or limit the high seas freedom of other States and their nationals to lay and maintain international cables is inconsistent with UNCLOS. If enforced, it would engender active challenges of the States of other non-OSPAR cable co-owners and flag States. The cables that currently are laid in the OSPAR high seas MPAs land in non-OSPAR State parties. The consortium ownership of the cables includes many companies that are legally domiciled in non-OSPAR State parties. The legal basis for OSPAR to impose restrictions directly or indirectly through its member States on non-OSPAR State parties and their nationals on the high seas requires explanation by OSPAR – so far not provided – as it is inconsistent with UNCLOS.

Unusually, BEP has no provision to grandfather cable previously laid when the BEP was activated by OSPAR. By analogy, the only example of a restriction on existing high sea cables whose status was changed by international law is seen in Article 51 of UNCLOS. In that case cables laid on the high seas, whose physical location status was converted to archipelagic waters, provided that existing cables laid when UNCLOS entered into force, were grandfathered such that they continued to enjoy a special status based on their high seas freedom status prior to UNCLOS entering force. BEP contains no such provision. This omission is not equitable given the hundreds of millions of dollars invested in these cable systems and the reliance on them by all nations. As noted by the UK Government, telecommunication cables are 'socially and economically crucial to the UK'[49] and according to the UN '[s]ubmarine cables are recognized as

---

47 Articles 87(c) and 112 of UNCLOS.

48 Articles 58 and 78 of UNCLOS.

49 Department of Environment, Food and Rural Affairs, 'UK marine policy statement published', 2011, <http://www.defra.gov.uk/news/2011/03/18/marine-policy-statement/> accessed 1 July 2017.

vitally important to the global economy and hence to economic growth.'[50] Any OSPAR regulatory scheme now or in the future involving international cables should include suitable grandfather provisions for international cables laid outside of territorial seas.

An EIA (or equivalent) should not be required before cable repairs take place given the importance of ensuring that emergency repairs are done as expeditiously as possible.[51] Nor are such reports required under UNCLOS, because historically there are no reasonable grounds to believe that laying or repairing a fibre optic cable may cause substantial pollution of or significant and harmful changes to the marine environment.[52] In the ABNJ, repairs to international cables, all surface laid, are rare, on average only about four repairs worldwide on the high seas per year.[53] EIAs for repairs in other waters, if required by a coastal State, under Article 206, should be part of the cable landing license process.

The unilateral imposition of MPAs in the ABNJ, if enforced directly or indirectly by OSPAR or State Parties of OSPAR against nationals of non-OSPAR States that own cables or cable ships, is inconsistent with UNCLOS.[54] First, no notice was provided to either these States or their national cable and cable ship owners. Second, there was no attempt by OSPAR or its members to exercise minimally acceptable compliance with the 'due regard' provisions of UNCLOS.[55] While UNCLOS does not accord rights to private companies that own and operate cables and cable ships, these companies do have recourse to their States who do have the ability to challenge these actions on their behalf under the provisions of UNCLOS.

---

50 UNSG Report A/70/74 (30 March 2015).

51 D. Burnett, D. Freestone, and T. Davenport, 'Submarine Cables in the Sargasso Sea: Legal and Environmental Issues in Areas Beyond National Jurisdiction', Report from the Workshop held 23 October 2014 at George Washington Law School, Washington, D.C. (16 January 2015), <https://www.iscpc.org/documents/?id=1792> accessed 1 July 2017.

52 Article 206 of UNCLOS. See ANNEX II.

53 ICPC, 'Annex B: Submarine Cable Repairs Worldwide 2008–2015', in *Submarine Cables and BBNJ*, 2016, <https://de.scribd.com/document/324478553/International-Cable-Protection-Committee-Submarine-Cables-and-BBNJ-White-Paper-Aug-2016> accessed 1 July 2017: Average number of repairs per annum with one or more cable faults per year on average by Territorial Waters/Exclusive Economic Zone.

54 See D. Burnett, R. Beckman and T. Davenport (eds.), *Submarine Cables: The Handbook of Law and Policy*, Martinus Nijhoff Publishers, 2014, pp. 208–212.

55 *Chagos Marine Protected Area* (Mauritius vs. United Kingdom), Arbitration Award, 2015, para. 515–518.

4 Current International Governance for Submarine Cables is Adequate

A salient aspect of the United Nations General Assembly Resolution 69/292 mandate is that any proposed terms in a possible new implementing agreement must not undermine the provisions of UNCLOS. In the particular case of international submarine cables this aspect is vital, based on their low impact to the marine environment and historically proven practicality that has provided the world with its amazing critical infrastructure communication system. The success of the world's submarine cable systems would not have occurred but for the crucial support it has received from UNCLOS, which comprehensively addresses rights and obligations of submarine cables in all of the maritime zones established by UNCLOS. Under UNCLOS, the freedom to lay cables includes those operations associated with this freedom such as cable route surveys and repairs.[56]

The provisions of UNCLOS on submarine cables are based on a Convention adopted in 1884 for the Protection of Submarine Telegraph Cables.[57] The essential principles have remained unchanged since the 19th century. This shows the stability and solidity of the legal framework. The long-standing regime has proven its ability to protect and foster the development of submarine telecommunication technologies, while preserving the marine environment. The protection of BBNJ is just one more challenge to be tackled in accordance with the relevant provisions of UNCLOS.

Submarine cables in the BBNJ are seldom disturbed once laid with no burial below the seabed surface. Worldwide cable fault records for the period of 2008–2015 show that in the BBNJ there are on average little more than four faults annually spread out in all of the world's high seas areas.[58] These few faults are generally associated with underwater landslides, seismic events, or other natural phenomena since there are few, if any, human generated events at those depths.

There is no requirement for repair permits on the high seas. While permits for emergency repairs to international cables in the EEZ are inconsistent with UNCLOS freedom to lay and maintain cables, several coastal States insist on

56 Articles 58(1) and 78(2) of UNCLOS.

57 Convention for the Protection of Submarine Telegraph Cables (adopted 14 March 1884, entered into force 1 May 1888) 24 Stat. 989, T.S. No 380 (1888).

58 See ICPC, *Submarine Cables and BBNJ*, 2016, <https://de.scribd.com/document/324478553/International-Cable-Protection-Committee-Submarine-Cables-and-BBNJ-White-Paper-Aug-2016> accessed 1 July 2017.

permits and delays with the negative results for the resiliency of the world's undersea cable communications. These facts underscore the unpredictable patchwork of a minority group of coastal States permitting regimes in their EEZ and the corresponding alarming delay to emergency cable repairs.[59] These coastal State regulations and actions are classified as excessive maritime claims.[60] Similar excessive permitting would not be helpful to introduce in BBNJ; it reinforces the wisdom in UNCLOS that allows for the freedom to lay and repair cables on the high seas proper.

4.1 *The Factual Context for International Law for Cables in the BBNJ*

These time tested and proven norms of international law codified in UNCLOS are fundamental to international submarine communications. To understand how well the current UNCLOS provisions work and the risks involved of unintended consequences in altering these balanced provisions, it is important to understand the practical reality about international telecommunication submarine cables.

First, the foremost priority for the submarine cable community is the integrity and resilience of the submarine cable systems, which are critical for a wide variety of essential services that we take for granted. From the submarine cable community perspective, additional regulations that undermine traditional freedom to lay and repair cables in areas beyond national jurisdiction would have a detrimental effect on the reliability of the cable network. For context in submarine telecommunication cable reliability and resilience, ocean policy makers and diplomats should consider seven key points that apply universally:

- As established in the prior section of this paper, cables have a neutral environmental footprint on the seabed. In the BBNJ, submarine cables are surface laid on the flat seabed, not buried; to avoid damage to potential biological 'hot spots' they are not laid on the tops or flanks of seamounts and avoid areas of active volcanism.
- There is no single global submarine cable network any more than there is a single world airline network. (There are about 236 active separate and decentralized international cable systems totalling 997,336 km.)
- Cable systems are either owned by consortia of four to thirty private companies or in some cases by a single company. About 99% of international

59 Ibid.

60 A. Roach and R. Smith, *Excessive Maritime Claims*, 3rd Ed., Martinus Nijhoff Publishers, 2012, Chapter 16, pp. 461–462, citing various States.

telecommunication cables are non-government owned. Cable systems are not 'flagged' to any one State.

- Cable repair arrangements are organized regionally by private contract – not by government mandate. Contracts require repair ships to sail within 24 hours of notice of a cable fault; goal: fast response and repair.
- There are about 59 cable ships in the world; about half are on stand-by to carry out emergency repairs pursuant to cable ship pooling contracts with various cable owners and cable ship operators, and the other half is laying new cables or performing other tasks (training, vessel maintenance, out of service cable recovery).
- Cable ships are expensive, custom built, conspicuous, require specialized crews, and fly diverse flags (UK, France, Marshall Islands, Singapore, Japan, China, Korea, UAE, Panama, Denmark, Norway, Spain, Italy, Philippines, Mauritius, Barbados, Belize, Indonesia). To be considered: competitive rates and efficiency.
- Cable repairs are urgent not only to restore service, but because each cable acts as the backup for other cables that are damaged and awaiting repair (resiliency).

In light of the above points, the most pressing concern for the cable community is the possibility that the existing UNCLOS provisions for submarine cables will be changed or overridden by a possible new environmental regulatory regime implemented under the BBNJ process or coastal State encroachment. The submarine cable community believes, given the critical importance of telecommunication cables, that the submarine cable provisions in UNCLOS should not be modified or subjected to any new regulatory burden associated with any new BBNJ implementing agreement or coastal State encroachment.

4.2 *Balancing Submarine Cables and Other Uses with ABNJ*

Submarine cables have long standing working relationships with other activities in ABNJ.

4.2.1 Merchant and Fishing Vessels and Pipeline Crossings

The submarine cable community understands that the freedom to lay and maintain cables on the high seas is not unqualified and has never treated it as such. Always present are the obligations to avoid conduct that prejudices the repair of other cables or pipelines,[61] to indemnify damage to any first laid cable

61 Article 112(2) of UNCLOS.

or pipeline that is crossed ('the first laid rule'),[62] to indemnify mariners or vessel owners who, through no fault of their own foul a cable, but sacrifice their gear to avoid injury to the cable,[63] and to show 'due regard' for the interests of other States in the exercise of the freedom of the high seas and with respect to activities in the Area.[64]

Unlike the high seas, in national waters, the most significant largest cause of faults – about 72–86% – comes from bottom trawl and similar aggressive fishing activities and contact with ship anchors.[65] Based on the structure provided in UNCLOS, over the past 166 years the submarine cable community has developed sound practices with the fishing and shipping industries including charting, education and liaison, and other time tested techniques that allow these risks to be managed and reduced.[66] In those cases, appropriate, domestic legislation and legal remedies in national admiralty courts[67] are adequate to provide a deterrent to culpably negligent or wilful conduct (excluding terrorism)[68] that threatens or damages the critical international submarine cable infrastructure.

It is emphasized that injury to a telecommunications cable results in zero marine pollution, only a disruption in communications.[69] With respect to crossing other cables or pipelines anywhere, the custom and practice of the

62 Article 115 of UNCLOS.

63 Article 114 of UNCLOS.

64 Article 87(2) of UNCLOS.

65 D. Burnett, R. Beckman and T. Davenport (eds.), *Submarine Cables: The Handbook of Law and Policy*, Martinus Nijhoff Publishers, 2014, p. 256.

66 R. Wargo, *Protecting Submarine Cables from Competing Uses*, in ibid., Chapter 11, pp. 254–279; ICPC, 'Actions for Effective Cable Protection (Post Installation)', Recommendation No. 6, Issue 9, 2015.

67 *Agincourt Steamship Company Ltd vs. Eastern Extension Australia and China Telegraphy Company Ltd*, 2 KB 305, 1907, United Kingdom; *Alex Pleven* (France), *Whiteman, Digest of International Law*, vol. 9, 1961, pp. 948–951; *American Tel & Tel Co vs. M/V Cape Fear* 763 F Supp. 97 (DNJ 1991), United States; *Peracomo et al vs. Sociéte Telus Communications, Hydro Québec, Bell Canada vs. Royal and Sun Alliance Insurance Company of Canada*, 2012 FCA 199, 2012, *aff'd* 2011 FC 494, 2011, Canada; *The Government of the Netherlands, Post Office vs. GT Manneteje-Van Dam [Fishing Cutter GO 4]*, File No 325/78, 1978, *aff'd sub nom G't Mannethe-Post Office*, File No 69 R/81 and File No 325/78, 1983, the Netherlands.

68 R. Beckman, 'Protecting Submarine Cables from Intentional Damage – The Security Gap', in D. Burnett, R. Beckman and T. Davenport (eds.), *Submarine Cables: The Handbook of Law and Policy*, Martinus Nijhoff Publishers, 2014, pp. 281–300.

69 M. Nordquist et al., *United Nations Convention on the Law of the Sea: A Commentary*, Volume II, Martinus Nijhoff Publishers, 1995, p. 914, No. 8; and Art. 79(3) of UNCLOS; a coastal State is allowed to delineate a pipeline route, but not a cable route.

industries involved allows for these events to take place routinely, safely, and in almost all cases without conflict. The engineers for the crossing systems meet and work out a crossing arrangement or in some case a formal agreement that complies with the due regard obligations in UNCLOS.[70] As with fishing and shipping, the current practices and protections for cable and pipeline crossings provided in UNCLOS are adequate and need no additional supplemental treaty provisions or super regulator. Nor is there a need with respect to cables for marine spatial planning since conflicts are historically well managed by those involved.

Instead of a new submarine cable treaty, provisions, or centralized marine spatial planning by a new or existing entity, greater compliance by States with their existing, but largely unfulfilled obligations under Article 113, would strengthen and enhance the reliability of the world's critical ocean infrastructure. The United Nations Omnibus Resolution on Oceans and the Law of the Sea[71] underscores this point:

> *Recognizing* that fibre optic submarine cables transmit most of the world's data and communications and, hence, are vitally important to the global economy and the national security of all States, conscious that these cables are susceptible to intentional and accidental damage from shipping and other activities, and that the maintenance, including the repair, of these cables is important, noting that these matters have been brought to the attention of States at various workshops and seminars, and conscious of the need for States to adopt national laws and regulations to protect submarine cables and render their wilful damage or damage by culpable negligence punishable offences, [...]
>
> 158. *Calls upon* States to take measures to protect fibre optic submarine cables and to fully address issues relating to these cables, in accordance with international law, as reflected in the Convention;
>
> 159. *Encourages* greater dialogue and cooperation through workshops and seminars among States and the relevant regional and global organizations on the protection and maintenance of fibre optic submarine cables to promote the security of such critical communications infrastructure;
>
> 160. *Also Encourages* the adoption by States of laws and regulations addressing the breaking or injury of submarine cables or pipelines

70 ICPC, 'Telecommunication Cable and Oil Pipeline/Power Cable Crossing Criteria', Recommendation No. 9A.

71 UNGA Res. 70/235 (23 December 2015).

> beneath the high seas done wilfully or through culpable negligence by a ship flying its flag or by a person subject to its jurisdiction, in accordance with international law, as reflected in the Convention;
>
> 161. *Affirms* the importance of the maintenance, including the repair, of submarine cables, undertaken in conformity with international law, as reflected in the Convention [...].

The resolution does not encourage or call upon States to change the provisions in UNCLOS that are involved with submarine cables, but simply to comply with the existing ones.

4.2.2 Deep Seabed Mining

The freedom to lay and maintain cables is further qualified by the obligation of the submarine cable community to exercise this freedom 'with due regard for this Convention with respect to activities in the Area.'[72] In this regard, the International Seabed Authority (ISA) has issued a technical study that directly confirms the common obligation of cable owners, the mining contractors and the ISA to provide notice and meaningful consultation among themselves before initiating their activities.[73] Since 2010, the ISA and the ICPC have productively worked together under a Memorandum of Understanding (MOU) to address a practical 'due regard' process.

While no mining exploitation licenses have been issued by the ISA, two concession areas have been identified with active submarine cable systems present.[74] Based on the joint workshop approach developed by the ISA and the ICPC, there is a high degree of confidence that any conflicts with these systems or future mining operations and other cable systems will be professionally resolved by the participants using the existing applicable provisions of UNCLOS and the custom and practice of the submarine cable community in similar crossing situations found in national waters. Accordingly, there is no need for any new implementing treaty to address issues or spatial planning about deep seabed mining and submarine cables in ABNJ. The current provisions in UNCLOS are adequate and well understood by those involved.

72 Article 87(2) of UNCLOS.

73 International Seabed Authority (ISA), 'Submarine Cables and Deep Seabed Mining. Advancing Common Interests and Addressing UNCLOS "Due Regard" Obligations', *Technical Study: No. 14*, 2015, p. 25, para. 4.

74 Ibid., p. 10, para 4, pp. 39 (Annex B), 41 (Annex C).

4.3 *Legal Environmental Aspects of Submarine Cables with BBNJ*

The current UNCLOS balance between submarine cables and the environment and sustainability is easily jeopardized by a potential new regulatory regime characterized by some as an enhanced 'conditional freedom of the seas'. But enhancing the freedom to lay cables with new conditions impedes the nimble, efficient, and innovative character of the submarine cable community in a spiralling tangle of what appears to be an unnecessary, ever changing, unpredictable and excessive BBNJ regulation imposed by well-intentioned new or existing regulatory entities.

The high risk of unintended consequences of such actions needs to be carefully considered based on the unique known situation of submarine cables in the marine environment. For example, even the indirect centralization and well-intentioned control of the world's submarine cable systems by a BBNJ EIA process and similar permission requirements by a new or existing governance entity raises troubling and unpredictable consequences that may well diminish the ability to counter cybersecurity threats to the world's decentralized undersea communication systems, making them more vulnerable to such threats. If a new or existing international entity is tasked with coordinating marine spatial planning for everything including cables, it will likely obtain all the data on the existing submarine cable positons and centralize them in a single data base. Terrorists or anarchists could then hack into a single target data base, dramatically increasing the risk that current cable systems will be compromised in some form.

A new BBNJ implementing regulatory approach for governance of submarine cables is not necessary for several reasons. Submarine cables are already well regulated by coastal States through national legislation whenever an international submarine cable land in the territory of a State or transits its territorial sea. The world's small fleet of cable ships is regulated by their Flag States and subject to port and coastal State jurisdiction as well. However, while non-flag State regulatory regimes do not apply on the high seas proper, it is the historical and scientific environmental record for modern submarine cables that speaks for itself. There is no justification in grafting either a new regulatory and governance regime on a proven ongoing process under UNCLOS or the creation of a super-international regulator for submarine cables when no significant harm exists.

The physical footprint of a submarine cable in the BBNJ area is only 17–22 mm in diameter – i.e. fits in the space between these parentheses ( ). Using a fibre optic cable diameter of 22 mm, assuming total fibre optic cable in service in the BBNJ area totals approximately 314,350 km, the total ocean coverage by cables is estimated to be about 6.9 km^2. Using a BBNJ area estimate of about 230,000,000 km^2, the percent of BBNJ area covered by in-service cables is about 0.00002 %. The amount of power in a submarine telecom cable is a constant DC current of about 0.6 to 1.0 amperes. By comparison, a laptop computer operates on about three amperes and most household circuit breakers are around 10–20 amperes.[75]

In terms of numbers of international cables, the cumulative impact is also minimal. The numbers of these dispersed garden hose-like structures in the vastness of the oceans are regionally small, stable, or experiencing only small increases in numbers to respond to increased capacity demand. One reason for the number stability, besides the considerable cost of a new cable system, is the new low cost upgrades that became available around 2006. These upgrades based on further division of the light spectrum, allow the capacity of an existing operational cable to be upgraded by large multiples by simply changing equipment at the cable landing stations, leaving the physical cable on the seabed undisturbed.[76]

Submarine cables are a lawful use of the sea and have now been in the world's oceans for over 166 years, from the telegraph era (1850–1950), to the telephone era (1950–1986) to the optical era (1987 to the present).[77] In this span of history, submarine cables have never been associated with the irreversible loss of any species.[78] In fact, out-of-service cables are used as artificial reefs and reused for monitoring of the ocean environment.[79]

---

75 D. Burnett, D. Freestone, and T. Davenport, 'Submarine Cables in the Sargasso Sea: Legal and Environmental Issues in Areas Beyond National Jurisdiction', Report from the Workshop held 23 October 2014 at George Washington Law School, Washington, D.C. (16 January 2015), <https://www.iscpc.org/documents/?id=1792> accessed 1 July 2017, p. 20, para 49.

76 B. Clesca, H. Fevrier and J. Schwartz, 'Upgrading Cable Systems? More Possibilities That You Originally Think Of', *Submarine Telecoms Forum*, vol. 66, November 2012, pp. 23–27.

77 S. Ash, 'The Development of Submarine Cables', in D. Burnett, R. Beckman and T. Davenport (eds.), *Submarine Cables: The Handbook of Law and Policy*, Martinus Nijhoff Publishers, 2014, pp. 19–40, and Appendix One (Timeline of the Submarine Cable Industry].

78 D. Smith, J. Suárez and T. Agardy, *Routledge Handbook of Ocean Resources and Management*, 2015, p. 353.

79 D. Burnett, R. Beckman and T. Davenport (eds.), *Submarine Cables: The Handbook of Law and Policy*, Martinus Nijhoff Publishers, 2014, pp. 214–215: 'Out-of-service Submarine Cables'.

The Oceans and the Law of the Sea Report of the Secretary-General[80] succinctly sums up the conventional wisdom about international submarine cables and the marine environment:

> 54. The environmental dimension of submarine cables is, however, less apparent. Submarine cables themselves are considered to have a low-carbon footprint and a small relative impact on the environment, with the maintenance of submarine cables causing the highest impacts as a result of the operation of the cable ships themselves.[81] Submarine cables have the potential to be an active contributor towards disaster warning and addressing climate change with work underway to examine the potential for monitoring purposes.[82]

The small impact of submarine cables on the environment is documented in a long record of collaborative study in well-established, peer reviewed international scientific journals, workshops, and studies with top scientists and legal scholars.

In 2009 UNEP, the World Conservation Monitoring Centre (WCMC) and the ICPC collaborated and published a cross-disciplinary review and study of the impact of submarine cables on the marine environment.[83] The 64 page UNEP-WCMC Report, based on a review of 191 cited peer reviewed scientific, academic, industrial and governmental studies and vetted by 18 external reviewers, concluded: 'as outlined in this report, the weight of evidence shows the environmental impact of fibre-optic cables is neutral to minor.' Since the UNEP-WCMC milestone report, approximately 25 other peer reviewed university and research institution studies have been completed on various aspects of submarine cables in the marine environment including leaching studies, seabed recovery studies, marine mammal and shark studies, and EMF.[84] The cumulative result of these studies echoes the UNEP-WCMC Report that modern submarine cables have a benign or neutral impact in the marine environment.

---

80 UNGA A/70/50 (30 March 2015).

81 C. Donovan, *Twenty Thousand Leagues Under the Sea: A Life Cycle Assessment of Fibre Optic Submarine Cable stems*, Degree Project, 2009.

82 See UNGA Res. 69/71 (2 December 2014), Add. 1; and UNGA Res. 69/79 (2 December 2014), Add. 1.

83 L. Carter et al., 'Submarine Cables and the Oceans: Connecting the World', *UNEP-WCMC Biodiversity*, Series No. 31, 2009, p. 54.

84 Many are listed in the bibliography section.

The low environmental foot print of submarine cable was reconfirmed in the standard treatise on submarine cables *Submarine Cables: The Handbook of Law and Policy* in 2014 where it is observed that in practice cables are laid to avert environmental harm by identifying fragile ecosystems that are bypassed during the cable route survey process.[85] A subsequent 2015 ISA publication *Submarine Cables and Deep Seabed Mining* notes that 'submarine cables have a reduced carbon footprint' and 'that their environmental impact is minor if not negligible.'[86] In 2015, an interdisciplinary workshop 'Submarine Cables in the Sargasso Sea: Legal and Environmental Issues in Areas beyond National Jurisdiction'[87] made by consensus several relevant BBNJ findings:

- For water depths over 2000 m, cables are laid directly on the seabed and hence seabed disturbance is minimal.
- The laying of a cable is intended to be a one-off operation in the 25-year design life of a cable, although faults may occur mainly via natural and human-related hazards. When repairs are needed, grapnels used for cable recovery may disturb the seabed along meter-wide paths. The recovered cable is repaired and lowered to the seabed to minimize further disturbance. Again, a repair is planned to be a one-off operation in a cables' remaining design life.
- Cable operations also have a low carbon footprint, and are done with concern for safety, fuel economy and the environment.
- Once the cable is laid, the physical impact on the seabed is minimal. The size of communications cables is small, ranging from 17 to 21 mm. Cables are protected by a substantial sheath of marine grade polyethylene which is inert in the ocean.

---

85 L. Carter, D. Burnett, and T. Davenport, 'The Relationship between Submarine Cables and the Marine Environment', in D. Burnett, R. Beckman and T. Davenport (eds.), *Submarine Cables: The Handbook of Law and Policy*, Martinus Nijhoff Publishers, 2014, pp. 179–212 (202); see also L. Carter, 'Submarine Cables and Natural Hazards', in ibid., pp. 237–254; L. Carter and A.H.A. Soons, 'Marine Scientific Research Cables', in ibid., pp. 323–338.

86 International Seabed Authority (ISA), 'Submarine Cables and Deep Seabed Mining. Advancing Common Interests and Addressing UNCLOS "Due Regard" Obligations', *Technical Study: No. 14*, 2015, p. 18 and Annex F (Environmental Impacts and Submarine Cables).

87 D. Burnett, D. Freestone, and T. Davenport, 'Submarine Cables in the Sargasso Sea: legal and environmental issues in areas beyond national jurisdiction', Report from the Workshop held 23 October 2014 at George Washington Law School, Washington, D.C. (16 January 2015), <https://www.iscpc.org/documents/?id=1792> accessed 1 July 2017.

- Research into cables and benthic organisms living on and in the seabed show that there is no statistical difference in the abundance and diversity for organisms living near and away from a cable.
- Studies have also been done on the direct environmental impact of cables on marine life[88] (including whales[89] and sharks[90]). Whale entanglements with cables ceased with the transition from telegraph to coaxial cables by the early 1960s which reflected improved cable design, laying techniques and seabed mapping.

While the Sargasso Sea workshop focused on the Atlantic Ocean ABNJ, the above findings are consistent with the industry custom and practice in other ABNJ areas in other oceans. These findings as well as the neutral nature of submarine cable impact to the marine environment were again confirmed in 2016 at an international workshop in Ankara, Turkey.[91]

The recent United Nations World Ocean Assessment reviewed submarine telecommunications cables and concluded that they 'have very limited environment impacts'.[92] It also acknowledges the socio-economic importance of cables and the role played by the ICPC in ensuring the safety of cables and reducing even further their minor impact on the environment.

Notwithstanding the thorough and consistent record of scientific and academic review of modern cables discussed above, a remark about the precautionary approach is appropriate. While the submarine cable service respects

---

88 See, for example, I. Kogan et al., 'ATOC/Pioneer Seamount cable after 8 years on the seafloor: Observations, environmental impact', *Continental Shelf Research 26*, 2006, pp. 771–787; B.M. Grannis, *Impacts of mobile fishing gear and a buried fibre-optic cable on soft-sediment benthic community structure*, MSc thesis, University of Maine, 2005; also E. Andrulewicz, D. Napierska and Z. Otremba, 'The environmental effects of the installation and functioning of the submarine SwePol Link HVDC transmission line: A case study of the Polish Marine Area of the Baltic Sea', *Journal of Sea Research*, vol. 49, 2003, pp. 337–345, that show no negative impact on the abundance and diversity of benthic organisms.

89 M.P. Wood and L. Carter, 'Whale Entanglements with Submarine Telecommunications Cables', *IEEE Journal of Oceanic Engineering*, vol. 33, 2008, pp. 445–450.

90 L.J. Marra, 'Shark bite on the SL Submarine Light Wave Cable System. History, Causes, and Resolution', *IEEE Journal of Oceanic Engineering*, 1989, pp. 230–237.

91 Workshop 'Legal Status of Submarine Cables, Pipelines and ABNJ', Ankara, Turkey, 7–8 April 2016, scheduled for publication, available at dehukam@ankara.edu.tr.

92 UNGA, 'The First Global Integrated Marine Assessment, World Ocean Assessment I', 2016, p. 6, <http://www.un.org/Depts/los/global_reporting/WOA_RPROC/Chapter_31.pdf> accessed 1 July 2017.

the precautionary approach when warranted, the unique status and cumulative studies of submarine cables make its application as a basis for new regulation of submarine cables in the BBNJ process inappropriate. The noted scientist Professor Sir Peter Gluckman, chief Science Advisor to the New Zealand government, observed:

> The Precautionary Principle was initially intended as a framework FOR ACTION in the face of scientific uncertainty – that is, not using an absence of evidence as reason not to act – for example on climate change. But Callon pointed out that, when applied to the innovation space, the Precautionary Principle was being wrongly framed as a reason for abstention and inaction. [...] The default position has insidiously shifted to an interpretation that allows nothing in the face of any uncertainty, which by definition must exist. And so the misuse of the principle has become a guiding tool for advocates trying to stop any particular innovation.[93]

There is a solid argument to carefully consider that innovation allowed by the existing UNCLOS provisions on the freedom to lay and maintain international cables is the core lifeblood of the modern internet. The *cloud* is in fact thriving because of the submarine cables that link data centre servers and peoples in all nations. Existing environmental requirements in UNCLOS are sufficient to safeguard the marine environment from the demonstrably nil to very small environmental risks posed by the small, chemically inert submarine cables.

The precautionary approach should not be mechanically applied to justify regulations of well-established routine international cable operations and routes. As documented in the many peer reviewed references in this paper, cable operations are known activities with a long historical record of safe interaction with the marine environment. These routine operations have no significant impact on the marine environment and certainly not one that would justify application of the precautionary approach to them. A new and untried regulatory regime may stifle and suffocate the innovation that has given the world its critical international cable infrastructure. This could lead to unintended consequences such as limiting or prolonging the efforts to bring fibre-optic cables to small islands and developing States or providing alternate cable routes and restoration options to minimize risks from natural disasters. Such consequences, delays in repairs, delays in connecting small island nations as

93 P. Gluckman, 'The Place of Science in Environmental Policy and Law', The Salmon Lecture to the Resource Management Law Association, Wellington, New Zealand, 2 September 2015.

well as African nations, and providing alternate cable redundancy for security have already been the reported focus of international leaders at high levels of the Organization for Economic Co-operation and Development (OECD)[94] and the International Telecommunications Union (ITU).[95]

4.3.1 Environmental Impact Assessments for Submarine Cables

Under UNCLOS, submarine cables in the ABNJ area are already subject to Article 206 (Assessment of potential effects of activities). Under Article 206, a State that has reasonable grounds for believing a planned activity under their jurisdiction or control may cause substantial pollution or significant harmful changes to the marine environment may as far as practicable assess the potential harm of such activities on the marine environment.[96] Thus the flag State of the cable ship or a State whose nationals own or operate an international submarine cable already have authority to carry out an environmental impact assessment (EIA) if justified.

By definition, submarine cables are not 'pollution of the marine environment', nor can cables realistically cause such pollution. A modern fibre-optic cable is not a substance or energy likely to result in deleterious effects as harm to living resources and marine life.[97] As demonstrated previously, submarine cables also do not cause 'significant harmful changes'[98] to the marine environment. In view of the substantial scientific record, Article 206 has not been applied on the high seas. The point is, however, that it remains a legal obligation that is in force, respected by the submarine cable community, and always available to States to protect the marine environment.

---

94 OECD, 'Broadband and the Economy', Ministerial Background Report: DSTI/ICCP/IE(2007)3/FINAL, June 2008, pp. 12–13 (what happens when the cables fail), and pp. 32–33 (physical infrastructure of broadband and ICT-enabled trade in services).

95 The Broadband Commission for Digital Development, 'Broad Band: A Platform for Progress', 2011, p. 44, <www.st.nmfs.noaa.gov/commercial-fisheries/foreign-trade> accessed 1 July 2017: 'Although Africa has, at present, low penetration of broadband, new initiatives are taking place to improve connectivity. These include national plans for background infrastructure and new links via submarine cables.' See also p. 155: 'A key indicator for broadband development in Africa is the deployment of basic infrastructure, such as international fibre-optic cables. Many African governments have co-sponsored new cables of this type along the continent's east and west coasts, with the aim of improving broadband connectivity.'

96 Article 206 of UNCLOS.

97 Article 1(4) of UNCLOS.

98 Article 206 of UNCLOS.

In view of Article 206, there is no need to create a new overlapping EIA obligation in a BBNJ implementing agreement for submarine cable laying and repair. As the UNEP-WCMC report highlights:

> EIAs for cable operations are rare and are generally limited to a coastal State's territorial sea. The European Union EIA Directive currently does not explicitly impose an EIA requirement on cable-laying projects.[99]

The Sargasso Sea workshop in particular extensively looked at all aspects of cables on the high seas and their relationship to the marine environment and the consensus reached was 'that the impact of submarine cables and cable operations in the deep water of the Sargasso Sea would also be minimal', because cables were not buried or laid on seamounts, hazards posed by turbidity currents were slight, and hence repairs were infrequent.[100] Furthermore the workshop reached consensus that there was no baseline in the deep ocean by which to compare any change in the marine environment and no cost benefit analysis to justify the delays and costs associated with an EIA in ABNJ where there was no 'clear benefit of EIA's in such areas'.[101] 'It was also agreed that an EIA (or equivalent) should not be required before cable repairs take place in the Sargasso Sea given the importance of ensuring that repairs are done as expeditiously as possible.'[102] This is common sense. Where coastal States have required an EIA for submarine cables, the normal time frame varies from weeks to years to carry out the EIA and submit it to the government authority making the request.[103] In the case of an emergency cable repair, this would be like having the fire department carry out an EIA on a burning building before attempting to bring the fire under control and extinguish it. For laying a new cable system, new delays for a high seas or ABNJ EIA would threaten the

99 L. Carter et al., 'Submarine Cables and the Oceans: Connecting the World', in *UNEP-WCMC Biodiversity*, Series No. 31, 2009, p. 37.

100 D. Burnett, D. Freestone, and T. Davenport, 'Submarine Cables in the Sargasso Sea: Legal and Environmental Issues in Areas beyond National Jurisdiction', Report from the Workshop held 23 October 2014 at George Washington Law School, Washington, D.C. (16 January 2015), <https://www.iscpc.org/documents/?id=1792> accessed 1 July 2017.

101 Ibid., p. 22.

102 Ibid., p. 23.

103 L. Carter, D. Burnett, and T. Davenport, 'The Relationship between Submarine Cables and the Marine Environment', in D. Burnett, R. Beckman and T. Davenport (eds.), *Submarine Cables: The Handbook of Law and Policy*, Martinus Nijhoff Publishers, 2014, pp. 179–212 (200–201).

viability of a project that depends on being innovative and nimble to increasing broadband demand and at the same time compliant with a budget, financing, and project timeline. In view of the demonstrated lack of any significant harm from these activities, there is no need for such new regulatory and untested BBNJ burden on submarine cables.

Article 206 is sufficient to safeguard the marine environment in the case of submarine cables in BBNJ. As such there is no reasonable justification to add to submarine cable operations, especially to emergency repairs, a new obligation to carry out an EIA under any new undefined BBNJ regulatory regime.

4.3.2 Marine Protected Areas in the ABNJ

The submarine cable community's use of the term Marine Protected Area (MPA) in the BBNJ context includes proposals based on formal MPAS, Vulnerable Marine Ecosystems (VME), Particularly Sensitive Sea Areas (PSSA), or similar formulations that may be considered in the PrepCom process.

The submarine cable community's position on MPAs in the BBNJ is neutral so long as there is due regard for the freedom to lay and maintain submarine cables. Historically, submarine cables have co-existed in MPAs with no significant harm to the environment.[104] In fact, scientists have concluded that cable protection zones with the appropriate environmental attributes such as rocky reefs to encourage fish aggregation, can make ideal *de facto* MPAs.[105]

The majority of submarine cable routes follow the same 'tried and true' historic paths of earlier telegraph cables.[106] This reflects the proven low environmental impact and natural hazard risks historically experienced with submarine cable laying along these routes. As a result, new MPAs may well be considered over existing cable routes. The cable community does not see this as a problem as submarine cables and MPAs are not mutually exclusive by any means.

Table 11.1 underscores that submarine cables are mutually consistent with MPAs and have been so for years.

104 L. Carter, D. Burnett, and T. Davenport, 'The Relationship between Submarine Cables and the Marine Environment', in D. Burnett, R. Beckman and T. Davenport (eds.), *Submarine Cables: The Handbook of Law and Policy*, Martinus Nijhoff Publishers, 2014, pp. 179–212 (202–207); D. Smith, J. Suárez and T. Agardy, *Routledge Handbook of Ocean Resources and Management*, Routledge, 2015, p. 360.

105 L. Carter, D. Burnett, and T. Davenport, ibid., p. 207; L. Carter et al., 'Submarine Cables and the Oceans: Connecting the World', in *UNEP-WCMC Biodiversity*, Series No. 31, 2009, p. 37.

106 N. Starosielski, *The Undersea Network*, Duke University Press, 2015, p. 2.

TABLE 11.1 *Analysis of International Submarine Cables in ABNJ MPAs*[107]

Date description	ABNJ
Total number of cable systems in data base in ABNJ	150
Total cable systems in MPAs	22
Percent of cables that cross MPAs	15%
Total km of cables in ABNJ in data base	314,350 km
Total fibre-optic km in ABNJ MPAS	5,362 km
Percent of total km in MPAs	1.7%

These remedies already work in UNCLOS and should not be undermined by a new super-international regulator with remit over international cables.

5 Conclusion

The UNCLOS obligation of due regard is sufficient to ensure a harmonious relationship between submarine cables and any high seas MPA that may appear in the future.[108] The careful existing balance in ABNJ between the environment (Articles 192 and 206) and the freedom to lay and maintain international submarine cables (Articles 87, 112–115 and 297) has worked successfully since UNCLOS came into force. For these reasons, whatever the outcome of the BBNJ process for MPAs and EIAs will be, the existing structure of UNCLOS for submarine cables should not be undermined by changes to existing time tested practices that have served humankind so well or by subjecting cables

107 Based on comparison of MPA data base (Marine Conservation Institute, MPAtlas, Seattle, 2016, <http://www.mpatlas.org/explore/> accessed 1 July 2017) and commercial data base of Global Marine Systems Limited (GMSL). The ICPC gratefully acknowledges the assistance by MP Atlas and the Marine Conservation Institute as well as by GMLS. The commercial data base includes data supplied by Global Marine Systems Limited; copyright (2016) Global Marine Systems Limited. This data or information is provided on a reasonable endeavors basis and Global Marine Systems Limited does not guarantee its accuracy or warrant its fitness for any particular purpose. Such data or information has been reprinted with the permission of Global Marine Systems Limited.

108 Article 87(2) of UNCLOS.

to a new regulatory BBNJ regime that is not required to protect the marine environment from cables. Put another way, there is no need to fix something that is not broken.

In this regard, the OSPAR approach of unilaterally declaring high seas MPAs and proscribing practices for laying and operating international cables in ABNJ is harmful to the world's critical international submarine cable infrastructure and the many benefits associated with them. The lesson from the OSPAR experience is that shortcuts that omit stakeholder consultation before issuing recommendations or regulations are a poor choice. But the good news is that there is no reason why in the course of workshops or other forums joint efforts cannot take place to rebuild confidence and make voluntary mutual progress on everyone's common objectives. Certainly the cable community has made it clear to OSPAR that it is willing to partner where necessary to accomplish environmental goals.

The submarine cable community's current view is that the ongoing BBNJ process should not change or add conditions to the existing provisions in UNCLOS that deal with submarine cables and not impose any new and additional EIA and MPA requirements for cables in a new implementing agreement. There is precedent for this approach. The drafters of the 2001 Convention on the Protection of the Underwater Cultural Heritage,[109] after considering the historical record that submarine cables had never been a threat or damaged underwater cultural heritage, exempted submarine cables from this treaty.[110] Similarly, the compelling scientific evidence, the long positive track record of submarine cables in the marine environment, and the vital role of cables as critical international infrastructure merits a similar result in the BBNJ process.

Submarine cables, with their small footprint, positive contribution to reducing greenhouse gases, and well-studied neutral environmental impact, stand uniquely apart from other high impact uses of concern to the BBNJ area such as shipping, deep seabed mining, fishing, pipelines and energy. Submarine cables should be expressly excluded from any new implementing agreement or further attempts at coastal State encroachment.

109 Convention on the Protection of the Underwater Cultural Heritage (adopted 2 November 2001, entered into force 2 January 2009) 41 ILM 37 (2002).

110 D. Burnett, R. Beckman and T. Davenport (eds.), *Submarine Cables: The Handbook of Law and Policy*, Martinus Nijhoff Publishers, 2014, Chapter 8 (Out-of-Service Submarine Cables), p. 200.

Annex A

SQUIRE
PATTON BOGGS

Comparison of Representative Telecommunication Cable and HVDC Cable on the High Seas/ABNJ/BBNJ

FEATURE	COMMENT	FIBER-OPTIC TELECOM CABLE	HVDC CABLE
Diameter		17mm (.0.67 inches) (domestic garden hose size)	Between 12cm (4.28 inches) and 20 cm (7.86 inches), depending upon rating
Insulation	No liquid oil insulation in power cables	Polyethylene	XPLE or mass impregnated paper
Longest cable laid (km)	To date, HVDC cables only laid in EEZ while fiber-optic laid in EEZ, on the continental shelf and the high seas	9400km Single Segment TransPacific 11,036km Multi Segment TransPacific 25,000km Multi landing (Europe-Asia)	590km (NorNed HVDC cable)
Deepest water depth of cable lay (m)		9200m	1600 m
Weight (1 m section) kg		0.59kg (1.3 pounds)	50kg (110 pounds) to 60kg (132 pounds)
Number of cable ships	Cable ships laying each type of cable are different and not interchangeable	About 50	About 4 to 5
Length of segment between splices (km)	For fiber-optic, segment is where the optical amplifier is inserted into the cable at the factory For HVDC, segment varies by the weight the cable ship can hold and cable loaded in a segment per voyage (100km)	10km-60km	Cable usually manufactured in 30km lengths then factory jointed
Capacity cable ship (km)		About 7000km to 9000km of deep-sea cable	About 100 km
Universal joint kit (UJK) available	For fiber-optic cable, UJK allows any cable ship to joint cables. At sea about 14 hours required for telecom spice. For power cable, jointing is carried out only by factory technicians. At sea about 6 days required for HVDC splice	Yes	no
Burial	Fiber-optic cables laid in high seas/ABNJ/BBNJ are not buried. No HVDC cables currently exist on the high seas/ABNJ/BBNJ	no	NA
UNCLOS	Both types of cables enjoy the same rights and obligations	yes	yes

squirepattonboggs.com

FIGURE 11.1 *Comparison of Representative Telecommunication Cable and* HVDC *Cable on the High Seas/*ABNJ/BBNJ*.*

Annex B

Environmental Impacts

A substantial peer-reviewed literature shows conclusively that submarine telecommunication cables have nil to minimal impact on the marine benthic environment.[111] There is no record of OSPAR or the author of the BEP considering these papers prior to issuing the BEP. Here is a synopsis of the main findings:

Operations

- For water depths deeper than about 1500m, cables are laid directly on the seabed. There is no requirement for protective burial hence seabed disturbance is minimal.[112]
- Laying is planned to be a one-off operation in the 20–25 year design life of a cable, but faults may occur mainly via human-related and natural hazards.[113]
- When repairs are needed, grapnels are used for cable recovery and these can disturb the seabed along a metre-wide path. The recovered cable is repaired and lowered to the seabed to minimise further disturbance.[114] Again, a repair is planned to be a one-off-operation in a cables' remaining design life.
- Cables on the continental shelf, where water depths are typically less than 200m, may be buried for protection against bottom trawl fishing and ships' anchoring – the main causes of cable faults. Burial by plough or water jet will disturb the seabed along a narrow corridor of ~2–8m wide depending on the plough size. Disturbed seabed recovers especially in the presence of

111 Reviews and analyses of that literature can be found in D. Burnett, R. Beckman and T. Davenport (eds.), *Submarine Cables: The Handbook of Law and Policy*, Martinus Nijhoff Publishers, 2014; L. Carter et al., 'Submarine Cables and the Oceans: Connecting the World', in *UNEP-WCMC Biodiversity*, Series No. 31, 2009.

112 Ibid.

113 M.E. Kordahi, S. Shapiro and G. Lucas, 'Trends in Submarine Cable System Faults', Proceedings SubOptic, 2007; S. Drew, 'Submarine Cables and Other Maritime Activities', in UNEP and WCMC, 'Submarine Cables and the Oceans: Connecting the World', *UNEP-WCMC Biodiversity*, Series No. 31, 2009, pp. 43–48.

114 K. Ford-Ramsden and D. Burnett, 'Submarine Cable Repair and Maintenance', in D. Burnett, R. Beckman and T. Davenport (eds.), *Submarine Cables: The Handbook of Law and Policy*, Martinus Nijhoff Publishers, 2014, pp. 155–177.

ocean currents and waves with no long lasting impact on the biota.[115] Burial is generally a non-repetitive activity unless a cable requires repair.

Telecommunication Cables

- The size of a cable in water depths greater than 1500m is 17–21mm diameter, hence its physical foot print is small.[116] Such depths account for over 85% of the world ocean. On the continental shelf, cable dimensions can increase to up to 50mm diameter due to the addition of galvanized wire armour.[117]
- Cables are protected by a substantial sheath of marine-grade polyethylene, which is chemically inert in the ocean.[118]
- New armoured cables may leech zinc from the galvanized wire, but laboratory tests performed in closed tanks show the amounts emitted are typically less than 10 parts per million. These concentrations decline over 3 months in the laboratory, and reduce markedly in the ocean due to dilution. Finally, zinc is an essential element needed for marine organisms.[119]

---

115 E. Andrulewicz, D. Napierska and Z. Otremba, 'The Environmental Effects of the Installation and Functioning of the Submarine SwePol Link HVDC Transmission Line: A Case Study of the Polish Marine Area of the Baltic Sea', *Journal of Sea Research*, vol. 49, 2003, pp. 337–345; B.M. Grannis, *Impacts of Mobile Fishing Gear and a Buried Fibre-optic Cable on Soft-sediment Benthic Community Structure*, MSc thesis, University of Maine, 2005.

116 L. Hagadorn, 'Inside Submarine Cables in Submarine Cables and the Oceans – Connecting the World', in UNEP and WCMC, 'Submarine Cables and the Oceans: Connecting the World', *UNEP-WCMC Biodiversity*, Series No. 31, 2009, pp. 43–48; K. Ford-Ramsden and T. Davenport, 'The Manufacture and Laying of cables', in D. Burnett, R. Beckman and T. Davenport (eds.), *Submarine Cables: The Handbook of Law and Policy*, Martinus Nijhoff Publishers, 2014, pp. 124–154.

117 L. Hagadorn, ibid.

118 K. Collins, *Isle of Man Cable Study – Preliminary Material Environmental Impact Studies*, University of Southampton, 2007.

119 K. Collins, *Isle of Man Cable Study – Preliminary Material Environmental Impact Studies*, University of Southampton, 2007.

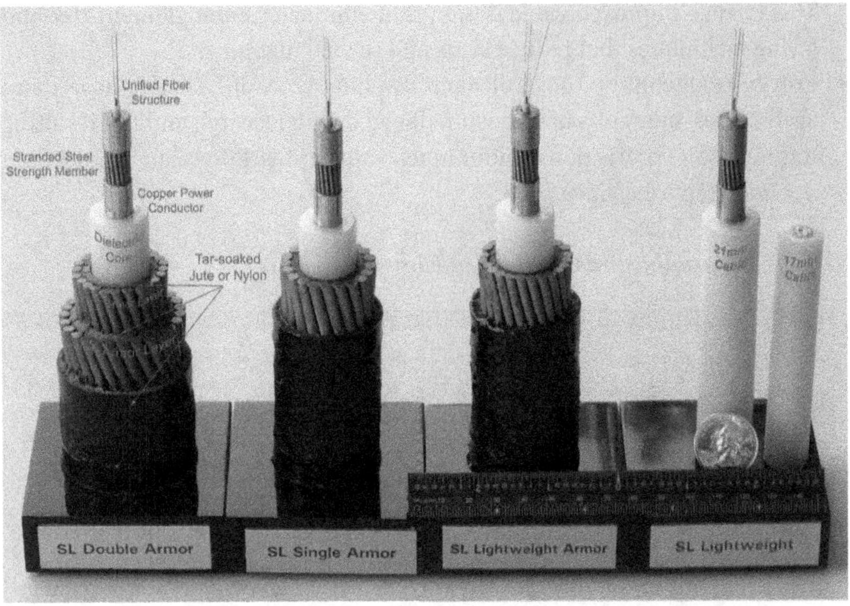

FIGURE 11.2 *A deep water sample of a fibre-optic cable, partly exposed to show the internal components, which include (from inside to out) glass fibres, copper conductor, steel wire strength member, polyethylene sheath (white) bite protection (silver) and black polyethylene external sheath.*

Interaction with Marine Biota

- Research into cables and benthic organisms show there is no statistical difference in the abundance and diversity for organisms living near and away from a cable.[120]
- Whale entanglements with cables ceased with the transition from telegraph to coaxial cables in the early 1960s and fibre-optic systems in the late 1980s.[121]

120 E. Andrulewicz, D. Napierska and Z. Otremba, 'The Environmental Effects of the Installation and Functioning of the Submarine SwePol Link HVDC Transmission Line: A Case Study of the Polish Marine Area of the Baltic Sea', *Journal of Sea Research*, vol. 49, 2003, pp. 337–345; B.M. Grannis, *Impacts of Mobile Fishing Gear and a Buried Fibre-optic cable on Soft-sediment Benthic Community Structure*, MSc thesis, University of Maine, 2005; I. Kogan et al., 'ATOC/Pioneer Seamount Cable after 8 Years on the Seafloor: Observations, Environmental Impact', *Continental Shelf Research*, vol. 26, 2006, pp. 771–787.

121 M.P. Wood and L. Carter, 'Whale Entanglements with Submarine Telecommunications Cables', *IEEE Journal of Oceanic Engineering*, vol. 33, 2008, pp. 445–450.

This change improved cable design that eliminates entanglement risk and laying techniques that reduce slack and seabed mapping.

- Fish bites, including those of deep dwelling crocodile sharks, have damaged cables, but faults have been reduced due to research and the resulting improved bite protection armour in relevant depths. Bite-related faults have not been reported since 2006.[122]

Telegraphic Era Coaxial Era Fibre Era

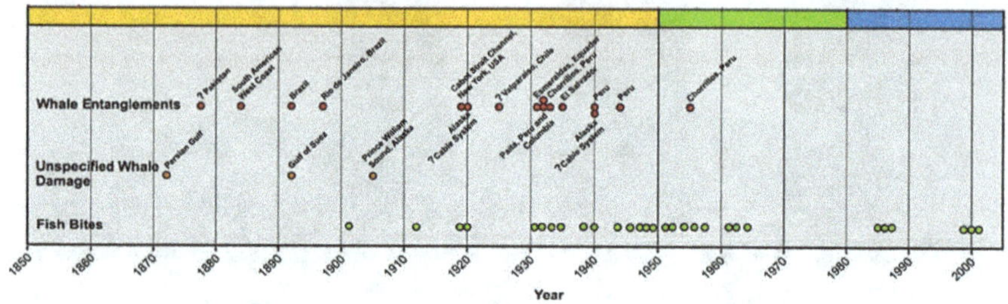

FIGURE 11.3 *Record of whale entanglements (red dots) that ceased in the 1950s with the introduction of improved cable design and deployment. The number of fish bites has declined with time due to improved bite protection. No cables have failed due to bites since 2006.*

Natural Hazards

Worldwide, natural hazards account for less than 10% of all cable faults.[123]

- Areas of elevated risk are generally associated with zones where tectonic plates converge such as the circum-Pacific rim. There, a combination of earthquakes, extreme weather and human modification of the land scape can generate cable-damaging submarine landslides and rapidly flowing

122 Ibid.; ICPC, 'Fish and Shark Bite Database', Report of the International Cable Protection Committee, 1988; L.J. Marra, 'Shark Bite with the SL Submarine Light Wave Cable System', *IEEE Journal of Oceanic Engineering*, vol. 14, 1989, pp. 230–237.

123 D. Burnett, R. Beckman and T. Davenport (eds.), *Submarine Cables: The Handbook of Law and Policy*, Martinus Nijhoff Publishers, 2014; M.E. Kordahi, S. Shapiro and G. Lucas, 'Trends in Submarine Cable System Faults', *Proceedings SubOptic*, 2007; S. Drew, 'Submarine Cables and other Maritime Activities', in UNEP and WCMC, 'Submarine Cables and the Oceans: Connecting the World', *UNEP-WCMC Biodiversity*, Series No. 31, 2009, pp. 43–48; M.P. Wood and L. Carter, 'Whale Entanglements with Submarine Telecommunications Cables', *IEEE Journal of Oceanic Engineering*, vol. 33, 2008, pp. 445–450.

turbidity currents (mud-laden currents).[124] For tectonically stable regions, the landslide/turbidity current risk is generally small.

- Sediment movement on the continental shelf may result in cable abrasion and fatigue but related faults are few attesting to effectiveness of cable armouring and/or burial. Climate change is strengthening ocean currents and the intensity/frequency of storms are likely to enhance sediment transport on the continental shelf 16, but burial and/or adequate cable armour will meet that challenge. Sea level rise and its amplification by storms may also pose a risk to coastal infrastructure.[125] The industry is well aware of these potential threats and has the expertise to adapt to change.

Submarine Power Cables

Much of the aforementioned summary applies directly to submarine power cables,[126] while noting that power cables are larger (150mm diameter in the case of Basslink DC interconnector and up to 260mm for HVAC power cables) than their telecommunications counterpart. In essence, power cables have a low environmental impact but questions have been raised concerning the impacts of EMF on the marine fauna. Reviews have not found conclusive evidence that EMF affects the behaviour of swimming or sessile organisms in the open ocean.[127] One of the most intensively studied cable systems is *Basslink*, which connects the power grids of Victoria and Tasmania. An independent board, commissioned by the Victoria and Tasmanian State governments and the Australian Federal government, concluded from the observational and modelling studies[128] that:

---

124 L. Carter et al., 'Insights into Submarine Geohazards from Breaks in Subsea Telecommunication Cables', *Oceanography*, vol. 27, 2014, pp. 58–67; L. Carter et al., 'Near-synchronous and Delayed Initiation of long run-out Submarine Sediment Flows from a record-breaking River Flood, Offshore Taiwan', *Geophysical Research Letters*, vol. 39, 2012.

125 Intergovernmental Panel on Climate Change, 'Observations: Oceans', in T. Stocker et al. (eds.), *Climate Change 2013: The Physical Science Basis, Contribution of Working Group 1 to the Fifth Assessment Report of the Intergovernmental Panel on Climate Change*, Cambridge University Press, 2013, pp. 255–315.

126 D. Burnett, R. Beckman and T. Davenport (eds.), *Submarine Cables: The Handbook of Law and Policy*, Martinus Nijhoff Publishers, 2014.

127 K. Ridgway and K. Hill., 'The East Australian Current', in E. S. Poloczanska et al. (eds.), *A Marine Climate Change Impacts and Adaptation Report Card for Australia*, 2009.

128 E.S. Poloczanska, A.J. Hobday and A.J. Richardson, *Marine Climate Change in Australia, Impacts and Adaptation Responses Report Card*, 2012; Normandeau Associates Inc. et al., *Effects of EMFs from Undersea Power Cables on Elasmobranchs and Other Marine Species*, published by US Dept. of the Interior, 2011.

- Magnetic fields associated with the cable are similar to those predicted from theory with effects dropping rapidly with distance from the cable. Beyond 20m distance, the field is indistinguishable from background fields.
- The Basslink cable was trenched by water jetting and after one year there was no evidence of the trench at over one third of observation sites. Elsewhere trench remnants were infilling with sediment that provided a habitat for a usually sparse benthic flora and fauna.
- Where the cable system crosses rocky reef, the cast iron pipe protecting the cable was encrusted with a biota that was indistinguishable from that of the natural reef.

Other Environmental Considerations

Submarine cables also play a role through the provision of data and knowledge on the marine environment – a benefit realized through strong and ongoing collaboration with the science community.

- Recovered cables yield biological samples for museum and university collections.[129]
- Cables underpin the communications and data transfer for major ocean science observatories including Ocean Networks Canada and the Ocean Observatories Initiative.[130]
- Cables act as sentinels of the deep ocean providing information on processes that shape the ocean floor such as submarine landslides and turbidity currents.[131]
- Discussions are underway between the industry, academia and the International Telecommunications Union regarding the feasibility of

---

129 J. Sherwood et al., 'Installation and Operational Effects of a HVDC Submarine Cable in a Continental Shelf Setting: Bass Strait, Australia', Journal of Ocean Engineering and Science, vol. 1, 2016, pp. 337–353.

130 BJAP, 'Final Panel Report: Basslink – Proposed Interconnector Linking the Tasmanian and Victorian Electricity Grids', Basslink Joint Advisory Panel, vol. 1, 2002; P.M. Ralph and D.F. Squires, 'The Extant Scleractinian Corals of New Zealand', Zoology Publications, Victoria University of Wellington, vol. 29, 1962, pp. 1–19.

131 L. Carter et al., 'Insights into Submarine Geohazards from Breaks in Subsea Telecommunication Cables', *Oceanography*, vol. 27, 2014, pp. 58–67; L. Carter et al., 'Near-synchronous and Delayed Initiation of long run-out Submarine Sediment Flows from a record-breaking River Flood, Offshore Taiwan', *Geophysical Research Letters*, vol. 39, 2012.

equipping cables with environmental sensors to monitor ocean change and hazards.[132]

- Cables provide a communications service that markedly reduces greenhouse gas emissions.[133] For a two-day teleconference between Stockholm and New York lasting eight hours per day, 5.7 kg of CO_2 would be released compared to 1920 kg emitted for the face-to-face meeting, which involved 16,000 km of air travel. This study highlights the modest carbon footprint of submarine telecommunications and their contribution to reducing greenhouse gas emissions.

Bibliography

Andrulewicz, E., Napierska, D. and Otremba, Z., 'The Environmental Effects of the Installation and Functioning of the Submarine SwePol Link HVDC Transmission Line: A Case Study of the Polish Marine Area of the Baltic Sea', *Journal of Sea Research*, vol. 49, 2003, pp. 337–345.

Beckman, R., 'Protecting Submarine Cables from Intentional Damage – The Security Gap', in D. Burnett, R. Beckman and T. Davenport (eds.), *Submarine Cables: The Handbook of Law and Policy*, Martinus Nijhoff Publishers, 2014, pp. 281–300.

Burnett, D., Beckman, R. and Davenport, T. (eds.), *Submarine Cables: The Handbook of Law and Policy*, Martinus Nijhoff Publishers, 2014.

Burnett, D., Beckman, R., and Davenport, T., 'Submarine Cables in the Sargasso Sea: Legal and Environmental Issues in Areas beyond National Jurisdiction', Report from the Workshop held 23 October 2014 at George Washington Law School, Washington, D.C. (16 January 2015), <https://www.iscpc.org/documents/?id=1792> accessed 1 July 2017.

Carter, L. et al., 'Submarine Cables and the Oceans: Connecting the World', in *UNEP-WCMC Biodiversity*, Series No. 31, 2009.

Carter, L. et al., 'Insights into Submarine Geohazards from Breaks in Subsea Telecommunication Cables', *Oceanography*, vol. 27, 2014, pp. 58–67.

Carter, L., Burnett, D., and Davenport, T., 'The Relationship between Submarine Cables and the Marine Environment', in D. Burnett, R. Beckman and T. Davenport (eds.),

132 Ocean Networks Canada, 2014, <http://www.oceannetworks.ca/about-us> accessed 1 July 2017.

133 Ocean Observatories Initiative, 2014, <http://oceanobservatories.org/> accessed 1 July 2017.

Submarine Cables: The Handbook of Law and Policy, Martinus Nijhoff Publishers, 2014, pp. 179–212.

Carter, L., 'Submarine Cables and Natural Hazards', in D. Burnett, R. Beckman and T. Davenport (eds.), *Submarine Cables: The Handbook of Law and Policy*, Martinus Nijhoff Publishers, 2014, pp. 237–254.

Carter, L. and Soons, A. H. A., 'Marine Scientific Research Cables', in D. Burnett, R. Beckman and T. Davenport (eds.), *Submarine Cables: The Handbook of Law and Policy*, Martinus Nijhoff Publishers, 2014, pp. 323–338.

Clesca, B., Fevrier and H., Schwartz, J., 'Upgrading Cable Systems? More Possibilities That You Originally Think Of', *Submarine Telecoms Forum*, vol. 66, 2012, pp. 23–27.

Davenport, T., *Science Technology and New Challenges to Ocean Law*, Brill, 2015.

Eccles, M., Ferencz, J. and Burnett, D., 'Submarine Power Cables', in D. Burnett, R. Beckman and T. Davenport (eds.), *Submarine Cables: The Handbook of Law and Policy*, Martinus Nijhoff Publishers, 2014, pp. 301–322.

Ford-Ramsden, K. and Burnett, D., 'Submarine Cable Repair and Maintenance', in D. Burnett, R. Beckman and T. Davenport (eds.), *Submarine Cables: The Handbook of Law and Policy*, Martinus Nijhoff Publishers, 2014, pp. 155–177.

International Seabed Authority (ISA), 'Submarine Cables and Deep Seabed Mining. Advancing Common Interests and Addressing UNCLOS "Due Regard" Obligations', *Technical Study: No. 14*, 2015.

Kogan, I. et al., 'ATOC/Pioneer Seamount Cable after 8 Years on the Seafloor: Observations, Environmental Impact', *Continental Shelf Research*, vol. 26, 2006, pp. 771–787.

Malphrus, S., 'Undersea Cables and International Telecommunications Resiliency', 34th Annual Law of the Sea Conference, University of Virginia, May 2010.

Manoj, C. et al., 'Can Undersea Voltage Measurements Detect Tsunamis?', *Earth Planets Space*, vol. 58, 2006, pp. 1–11.

Marra, L.J., 'Shark bite on the SL Submarine Light Wave Cable System. History, Causes, and Resolution', *IEEE Journal of Oceanic Engineering*, 1989, pp. 230–237.

Monastersky, R., 'The Next Wave', *Nature*, vol. 483, 2012, p. 144–146.

Müller, C. et al., 'Temperatures in Shallow Marine Sediments: Influence of Thermal Properties, Seasonal Forcing, and Man-made Heat Sources', *Journal of Applied Thermal Engineering*, vol. 108, 2016, pp. 20–29.

Nordquist, M. et al., *United Nations Convention on the Law of the Sea: A Commentary*, Volume II, Martinus Nijhoff Publishers, 1995.

Roach, A. and Smith, R., *Excessive Maritime Claims*, 3rd Ed., Martinus Nijhoff Publishers, 2012.

Smith, D., Suárez, J. and Agardy, T., *Routledge Handbook of Ocean Resources and Management*, Routledge, 2015

Starosielski, N., *The Undersea Network*, Duke University Press, 2015.

Wargo, R., *Protecting Submarine Cables from Competing Uses*, in D. Burnett, R. Beckman and T. Davenport (eds.), *Submarine Cables: The Handbook of Law and Policy*, Martinus Nijhoff Publishers, 2014, pp. 254–279.

Wood, M.P. and Carter, L., 'Whale Entanglements with Submarine Telecommunications Cables', *IEEE Journal of Oceanic Engineering*, vol. 33, 2008, pp. 445–450.

Worzyk, T., *Submarine Power Cables: Design, Installation, Repair and Environmental Aspects*, Springer, 2009.

Index

www.ingramcontent.com/pod-product-compliance
Lightning Source LLC
LaVergne TN
LVHW020634300926
849161LV00004B/700